Engineering Materials

This series provides topical information on innovative, structural and functional materials and composites with applications in optical, electrical, mechanical, civil, aeronautical, medical, bio- and nano-engineering. The individual volumes are complete, comprehensive monographs covering the structure, properties, manufacturing process and applications of these materials. This multidisciplinary series is devoted to professionals, students and all those interested in the latest developments in the Materials Science field, that look for a carefully selected collection of high quality review articles on their respective field of expertise.

Indexed at Compendex (2021) and Scopus (2022)

Muhammad Ikram · Ali Haider · Sawaira Moeen

Ferrite Nanomaterials for Wastewater Treatment

Springer

Muhammad Ikram
Solar Cell Applications Research Lab
Department of Physics
Government College University
Lahore, Pakistan

Ali Haider
Department of Clinical Sciences
Muhammad Nawaz Shareef University
of Agriculture
Multan, Pakistan

Sawaira Moeen
Solar Cell Applications Research Lab
Department of Physics
Government College University
Lahore, Pakistan

ISSN 1612-1317 ISSN 1868-1212 (electronic)
Engineering Materials
ISBN 978-3-032-09140-6 ISBN 978-3-032-09141-3 (eBook)
https://doi.org/10.1007/978-3-032-09141-3

This Springer imprint is published by the registered company Springer Nature Switzerland AG
The registered company address is: Gewerbestrasse 11, 6330 Cham, Switzerland

Dedicated to

My Parents
For their support throughout my education as well as for teaching me to be independent and responsible

My Brothers and Sisters
For their love and support

Preface

One of the most significant issues facing living things in the twenty-first century is the world's water supply. The fast growth of the population leads to a heightened need for drinkable water. The use of dyes in companies is perpetually rising, leading to the poisoning of groundwater through their effluents. The heightened use of chemical fertilizers and pesticides in agriculture has exacerbated environmental degradation. Consuming water containing chemicals elevates the chance of significant health problems, including hormonal disruption, cancer, renal and cardiovascular disorders, and aberrant brain function, among others. Furthermore, pollution adversely impacts the reproductive capacity of aquatic creatures. It diminishes the regenerative and reproductive abilities of fish and plants. Due to the consumption of contaminated water, animals are vulnerable to many ailments.

To overcome the above-mentioned challenges, several physical, chemical, and biological approaches were used to purify the wastewater. However, these techniques have disadvantages due to their expense, complex procedure, and production of insufficient byproducts. Among reported techniques, photocatalysis proved to be beneficial owing to its cost-effectiveness and environmental friendliness. This book helps students, emerging researchers, and scientists understand wastewater purification techniques. Firstly, ferrite nanomaterials with their classification were thoroughly explained. This book also provides a comprehensive analysis of the synthesis of ferrite nanomaterials using various top-down and bottom-up approaches. Structural, morphological, and magnetic properties of ferrite were briefly discussed. It will also help the reader to understand the chemical composition of cationic and anionic dyes. Lastly, the factors affecting the photocatalytic activity were thoroughly explained.

Lahore, Pakistan — Muhammad Ikram
Multan, Pakistan — Ali Haider
Lahore, Pakistan — Sawaira Moeen

Acknowledgments All praises to "ALLAH," the Almighty, most Gracious, the most Merciful, and the Sustainer of the worlds, who sent us "MUHAMMAD (PBUH)" as a blessing for whole universe and the best teacher with the ultimate source of wisdom "HOLY QURAN." First and foremost, we want to express our gratitude to ALLAH, the Almighty, for bestowing upon us the health to publish our book.

We express our sincere appreciation to Zernab Mateen, friend of author Sawaira Moeen, for her essential support in obtaining permits from publishers for the reuse of figures. Her diligent efforts in figure editing have significantly improved the quality of this work.

Finally, we acknowledge the unwavering support of our families and colleagues whose encouragement provided the foundation for this achievement. This book is a testament to their belief in our vision

Competing Interests The authors have no competing interests to declare that are relevant to the content of this manuscript.

Contents

About the Authors

Dr. Muhammad Ikram obtained his Ph.D. degree in Physics from the Department of Physics, Government College University (GCU) Lahore through the Pakistan-US joint project between the Department of Physics, GCU Lahore, Pakistan, and the University of Delaware, USA in 2015. He served as deputy director of Manuscript Science at Punjab textbook board (Pakistan). Later on (2017–to date), Ikram joined the Department of Physics, GC University Lahore as an Assistant Professor of Physics and Principal of Solar Cell Applications Lab and Advance Nanomaterials Research Lab. Ikram received the Seal of Excellence Marie Skłodowska Curie Actions Individual Fellowship in 2017 and 2020. In 2021–25, Ikram was included in the 2% top scientists from Pakistan announced by Stanford University. In 2024, he was awarded best publication award from HEC, Paksitan. His research interest involves the synthesis and characterization of inorganic semiconductor nanomaterials, 2D materials for water treatment optoelectronic, and electrocatalytic applications.

Dr. Ali Haider obtained his Ph.D. degree in Clinical Medicine from the Department of Veterinary Medicine (UVAS), Lahore in 2021. He served as Dairy Farm Manager at reputed commercial dairies of Punjab, (Pakistan). Later on (January, 2022–to date), Ali joined the Faculty of Veterinary and Animal Sciences, Muhammad Nawaz Shareef University of Agriculture Multan (MNS-UAM) as an Assistant Professor of Clinical Medicine. His research interest involves the Green synthesis and characterization of inorganic semiconductor nanomaterials, Polymeric nanostructures, 2D materials, wastewater treatment, antiparasitic and antibacterial evaluation against multiple drug resistant (MDR) pathogens.

Sawaira Moeen completed her M.Phil (Physics) from Government College University Lahore (Punjab, Pakistan) in 2025 under the supervision of Dr. Muhammad Ikram. Sawaira is currently working as a research assistant under NRPU project entitled "Self-supported 2D-Materials as an Efficient Electrocatalyst for Hydrogen Evolution Reaction" in Solar Cell Applications Research Lab and Advance Nanomaterials Research Lab. Sawaira has more than 30 international publications, including review papers with book chapters and book. Her research interest includes photovoltaics, synthesis of nanomaterials, electrochemical water splitting, energy storage devices, catalytic and photocatalytic degradation for wastewater treatment.

Abbreviations

AC	Activated carbon
AOPs	Advanced oxidation process
BG	Brilliant Green
CB	Conduction band
CR	Congo-red
CV	Crystal violet
e^-	Electron
EBT	Eriochrome Black T
Eg	Bandgap energy
ESR	Electron spin resonance
FNMs	Ferrite nanomaterials
Hc	Coercive force
HOMO	Highest occupied molecular orbital
LMG	Leucomalachite green
LUMO	Lowest unoccupied molecular orbital
MB	Methylene blue
MF	Magnetic field
MG	Malachite green
M-H	Magnetization hysteresis
MO	Methyl orange
Mr	Remanent magnetization
Ms	Saturation magnetization
MV	Methyl violet
NF	Nanofiltration
NMs	Nanomaterials
NPs	Nanoparticles
PC	Photocatalyst
PCA	Photocatalytic activity
RhB	Rhodamine B
RO	Reverse osmosis
SFs	Spinel ferrites

SiO_2	Silica
UF	Ultrafiltration
VB	Valance band
VSM	Vibrating sample magnetometer

Chapter 1
Ferrites: Introduction

A category of magnetic materials mostly composed of iron oxide (Fe_2O_3) is referred to as ferrites; the term ferrite originates from the Latin word for iron (Fe), "ferrum." Ferrites are frequently encountered in the natural world, exhibiting a dark brown or gray hue, and possess a mechanical characteristic that is notably hard but fragile [1]. Magnetic materials have been recognized by humanity since the early 20th century [2, 3]. Ferrite nanomaterials (FNMs) were identified as economically useful components by Snoek in 1947. The extensive manufacture of televisions in the 1950s catalyzed the expansion of the ferrite sector, thereby benefiting the national economy. Currently, FNMs have numerous uses across several sectors, including computers, memory cores, telecommunications, cars, etc. [4]. FNMs are compounds formed by the amalgamation of several bivalent oxides with trivalent iron oxide. The majority are effectively balanced, including bivalent oxide and ferrous oxide in a 1:1 ratio; however, some exhibit deviations [5]. FNMs exhibit poor conductivity, little dielectric loses, and high flexibility, making them significant prospects for usage in factories.

FNMs show pronounced ferrimagnetism due to the presence of metal electrons (e^-) and oxygen ions at lattice locations with opposing spin configurations. They exhibit net resultant magnetization as a consequence of uneven antiparallel spin moments [6]. Due to inherent atomic interactions among oxygen and metal ions, FNMs possess more resistance than ferromagnetic metals. Due to their diverse chemical arrangements, FNMs exhibit significant variations in magnetic characteristics, since their microscopic (atomic) structure consists of two or more electromagnetic sublattices. FNMs are non-conductive magnetic materials, resulting in reduced eddy current and ohmic losses [7].

M. Ikram et al., *Ferrite Nanomaterials for Wastewater Treatment*, Engineering Materials, https://doi.org/10.1007/978-3-032-09141-3_1

1.1 Classification of Ferrites

Ferrites are categorized based on two fundamental criteria: magnetic characteristics and crystal structure. The magnetic characteristics of FNMs arise from the mutual interactions and electronic configurations of the ions. Studying crystal structure is vital to comprehend the fundamental characteristics of FNMs. Ferrites are classified into four categories based on crystal structure: orthoferrites, garnet, spinel, and hexagonal, while they are further divided into hard and soft ferrites according to their magnetic properties (Fig. 1.1).

1.1.1 Spinel Ferrites (SFs)

There are several different divalent ions that are usually recognized as Fe^{2+}, cobalt (Co^{2+}), nickel (Ni^{2+}), cadmium (Cd^{2+}), zinc (Zn^{2+}), and manganese (Mn^{2+}). SFs have a standard chemical formula of MFe_2O_4, where M denotes divalent ions [9]. Normal, inverse, and complex spinals are the three forms of SFs that may be distinguished from one another based on their crystal structures (Fig. 1.2) [10]. In normal SFs, M(II) occupies a tetrahedral position, whereas Fe(III) occupies octahedral sites; $ZnFe_2O_4$ exemplifies an ordinary normal SF [11]. The spaces of octahedral sites are often greater than those of tetrahedral sites; hence, cations with lower radii tend to occupy M sites, whereas cations with larger size preferentially occupy Fe(III) positions. In inverse SFs structures, 50% of the Fe(III) occupies tetrahedral positions, whereas M(II) and the remaining Fe(III) are situated at octahedral sites, exemplified by $NiFe_0O_4$ [12]. In complex SFs, M(II) and iron are arbitrarily distributed across the tetrahedral and octahedral sites. $MnFe_2O_4$, as a depiction of complex SFs, may be properly expressed as $Mn_{0.8}Fe_{0.2}(Mn_{0.2}Fe_{1.8})O_4$ [13]. The initial use of SFs occurred using magnet Fe_3O_4 three millennia ago; nevertheless, additional research was absent for an extended period [14]. During the early 1990s, a systematic investigation of nanoscale SFs occurred alongside the fast advancement of nanomaterials. In recent years, scientists have focused more on the catalytic applications of nanoscale

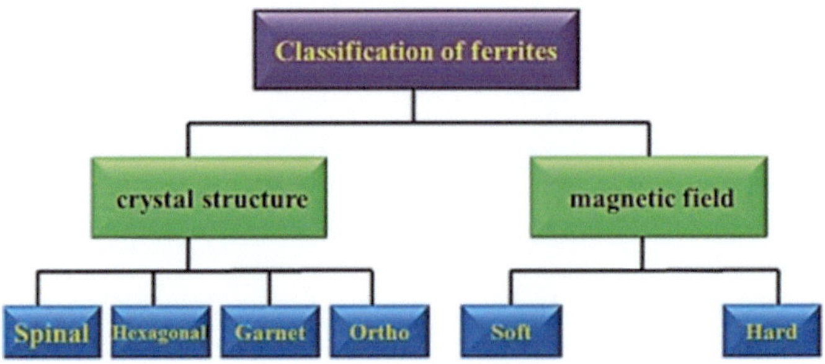

Fig. 1.1 Classification of FNMs. Reproduced with permission from [8]

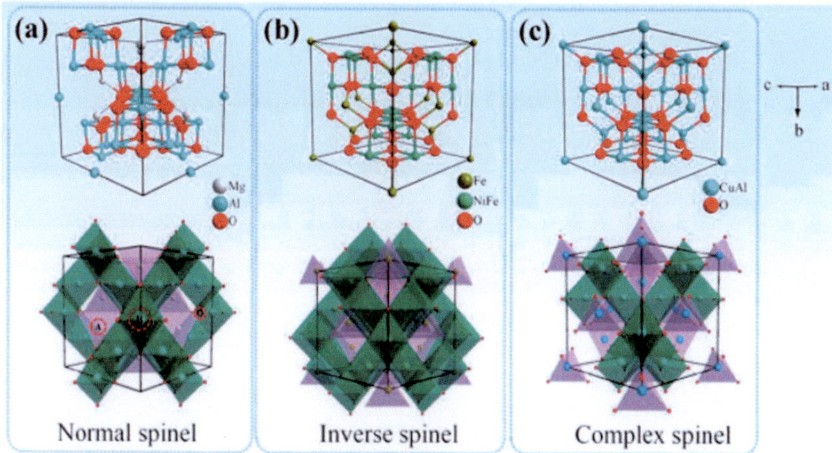

Fig. 1.2 Structure of **a** normal spinel, **b** inverse spinel, and **c** complex spinel. Reproduced with permission from [16]

SFs and their composite derivatives in the fields of energy and the natural world. In comparison to bulk materials, nanoparticles exhibit enhanced catalytic, physical, optical, and electrical capabilities. The extensive acid and basic adaptation range renders SFs advantageous for real-world use. Several SFs may be readily reused using an external magnet, aligning with the concepts associated with sustainable chemistry. The catalytic efficacy of pristine SFs is constrained by poor conductivity and inadequate electron movement; thus, numerous studies concentrate on investigating superior catalytic nanomaterials and their enhancement strategies to regulate particle dimension, anatomy, amalgamation, and surface structure [15, 16].

1.1.2 Hexagonal Ferrites

Hexagonal ferrites exhibiting a magneto-plumbite structure (space group ~ P63/mmc, 64 ions per unit cell) are well-established for their exceptional magnetic properties, rendering them suitable for microwave absorption at lower wavelengths and diverse applications [17]. One of these ions is a cation with a divalent form, commonly Ba^{2+}, Pb^{2+}, or Sr^{2+}, while the other 24 are Fe_3 cations and 38 are O_2 anions, allocated between five symmetrical sites: one bipyramidal, one tetrahedral, and three octahedral (Fig. 1.3) [18, 19]. These cations can then be substituted by a second similar-sized cation, leading to a solid mixture and alteration in their properties. Hexaferrites are magnetic Fe_2O_3 with a hexagonal phase and are crucial in the field of electronics because of their dielectric properties [20]. They were identified and defined in the early 1950s, and their economic significance has generated a substantial surge in interest since that time. They are amongst the most extensively studied substances

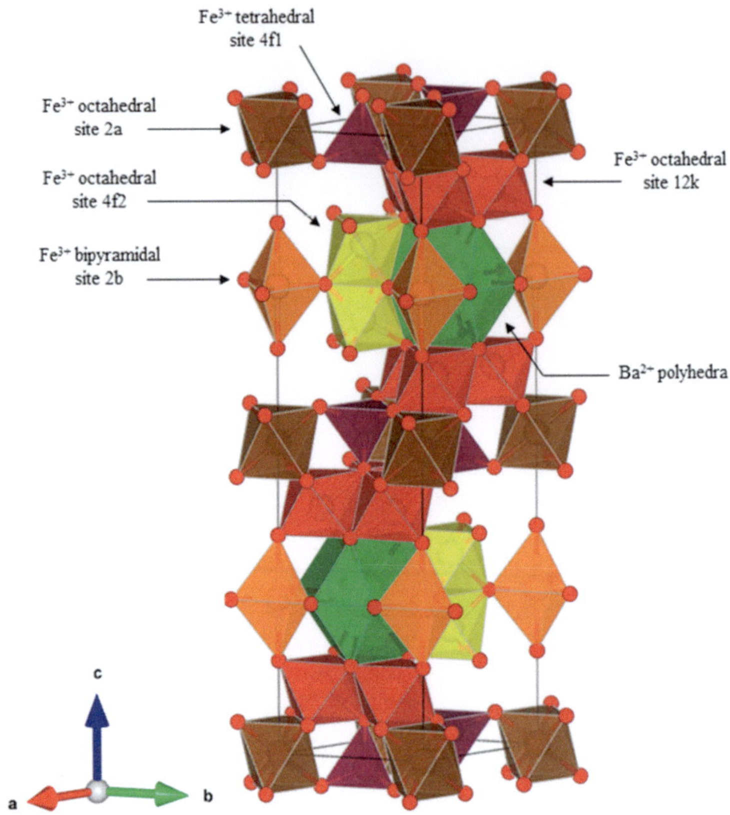

Fig. 1.3 Structure of hexagonal ferrite. Reproduced with permission from [22]

in high-frequency (lower wavelength) devices. They are optimal for purposes like radar absorption materials owing to their elevated flexibility, significant coercivity, robust chemical durability, commendable resistivity, and little eddy current losses. Several types of hexaferrites are M, X, Y, and Z types [21].

1.1.2.1 M-Type Hexaferrite

M-type is classified among ferrimagnetic oxides, characterized by the $AFe_{12}O_{19}$ formula, where A denotes divalent cations [23]. By Adelskold's investigation of $PbFe_{12}O_{19}$, M-type and its solid solution forms possess the same crystal phase (space group ~ P63/mmc, hexagonal unit cell: a = b = 5.90 Å and c = 23.30 Å) [24]. In terms of their crystal phases, these types of ferrites are composed of R and S blocks. The letter S represents a spinel, while the R represents hexagonal blocks. The unit cell is made up of RSR*S* components, where R* and S* have the identical atomic arrangements as R and S blocks, correspondingly, and rotate at an angle of 180°

around corresponding blocks [25, 26]. In their study, Alam et al. [27] found that the $BaZn_{0.5}Co_{0.5}ZrFe_{10}O_{19}$ exhibits a reflection loss (R_L) of -10 dB and an operating bandwidth that spans from 8 to 12 GHz. $BaCe_{0.05}Fe_{11.95}O_{19}$ was prepared by Chang et al. [28] by the utilization of the citrate sol–gel technique with R_L of -37.4 dB at 12.8 GHz. The influence of Sm^{3+} on the characteristics of these ferrites was investigated by Yasmin and colleagues [29]. Using SEM, the platelet-like particles with a hexagonal shape were identified, which are good for activities involving microwave absorption. With VSM, it was established that every sample has a coercivity that was greater than 1.3 kOe, evidence that these substances are hard magnetic.

1.1.2.2 W-Type Hexaferrite

W ferrites possess the formula $BaMe_2Fe_{16}O_{27}$, whereby Ba may be replaced by other alkaline earth metals, and Me typically denotes a transition metal. The inaugural documented W ferrite was Fe_2W ($BaFe_2Fe_{16}O_{27}$) [30], however, it was originally synthesized solely as a mixed phase with M and X ferrites. Fe_2W with mono phase exhibited magnetization along the c-axis (hardness ~ 5.5 GPa) of its hexagonal phase, while demonstrating greater conductivity (lower electrical resistance) than barium ferrite, attributed to the presence of Fe^{2+} ions [31].

1.1.2.3 X-Type Hexaferrite

The X ferrites were initially documented in 1952 as a combined state with M or W ferrite, with the $Ba_2Me_2Fe_{28}O_{46}$, where Me are transition metals. The initial documented X ferrite was Fe_2X, where Me = Fe^{2+}, and it exhibited unidirectional anisotropy along the c-axis (density ~ 5.29 g cm^{-3}) [32]. All X ferrites exhibit unidirectional anisotropy at ambient temperature, with the exception of Co_2X, having 2688 g mass, which displays a conical shape of magnetization (angle ~ 74°) to the c-axis [33]. Similar to the W ferrite it mimics, this cone reorients to align along the c-axis at an elevated temperature of 143 °C [34].

1.1.2.4 Y-Type Hexaferrite

The Y ferrites ($Ba_2Me_2Fe_{12}O_{22}$) were the inaugural ferroxplana FNMs identified, and it is now established that all these FNMs exhibit an optimal plane of magnetization orthogonal to the c-axis at ambient temperature [35]. The first discovered Y FNMs were Zn_2Y and Co_2Y (mass ~ 1410 g). Co_2Y exhibits transverse magnetic anisotropy at a typical temperature, which transitions to a conical magnetization below – 58 °C. The asymmetry persists in the optimal plane from this value to the Curie point. Cu_2Y is the sole Y ferrite identified to have a preferential unidirectional orientation of magnetization [36, 37].

1.1.2.5 Z-Type Hexaferrite

The Z ferrites ($Ba_3Me_2Fe_{24}O_{41}$) were identified concurrently with the ferroxplana Y ferrites. All Z ferrites exhibit unidirectional anisotropy aligned with the c-axis, except for Co_2Z (mass ~ 2522 g, density ~ 5.35 g cm^{-3}), which demonstrate planar anisotropy at ambient temperature and possesses a complicated magnetic anisotropy characterized by a minimum of four distinct anisotropic stages. At lower temperatures, Co_2Z exhibits conical magnetization to the c-axis (angle ~ 65°), which remains invariant down to −103 °C. Around this temperature and −53 °C, the angle grows to 90°, and the favored magnetization continues in the lowest plane until it transitions to the c-axis at a temperature around 207 and 242 °C [38].

1.1.3 *Garnet Ferrites*

Garnet ferrites ($X_3Fe_5O_{12}$, including two metal ions: one is a rare earth metal such as Sm, Gd, La, Ce, and Dy, and the other is Fe) are utilized in microwave, sound, and magneto-optical technologies. They are often recognized as minerals with dodecahedral (12-coordinated) sites alongside octahedral and tetrahedral sites. They possess magnetic hardness and have an orthorhombic crystal phase, characterized by an oxygen polyhedron encasing the cations. Trivalent cations, including rare earth elements (Sm, Gd, La, Ce, Dy, Er) and Fe3þ, occupy tetrahedral (d), octahedral (a), or dodecahedral (c) positions, the latter being a deformed 12-sided polyhedron. The connection among tetrahedral and octahedral sites is opposite one another, and the resultant magnetic moment is opposed to the rare earth ions located at the "c" sites. Their unit cell is cubic with an edge length of around 12.5 Å. The crystal phase of garnet is highly intricate, making it challenging to create a 2D depiction that clearly depicts all the ions within the unit cell. Figure 1.4 illustrates only an octant of the garnet framework, highlighting the locations of the cations. The garnet framework consists of octahedral locations (trivalent cations enclosed by 6 oxygen ions), tetrahedral sites (trivalent cations bounded by 4 oxygen ions), and dodecahedral sites (trivalent cations enclosed by 8 oxygen ions), with their orientations illustrated in Fig. 1.5. Yttrium iron garnet $Y_3Fe_5O_{12}$ is a recognized garnet [39–41].

1.1.4 *Ortho Ferrites*

Orthoferrites ($RFeO_3$, R represents rare-earth elements) have an orthorhombic phase (space group ~ Pbnm). These materials are classified as soft ferromagnets, including $DyFeO_3$, $PrFeO_3$, and $LaFeO_3$ [43]. These are cubic ferrites with a somewhat deformed perovskite structure. The phrase "perovskite structure" denotes the atomic configuration present in oxides characterized by the formula RMO_3, exemplified by $BaTiO_3$ and $PbTiO_3$. Orthoferrites are utilized in gas separating devices, dye degradation activity, spin valves, electrochemical H_2 generation, a network of communication

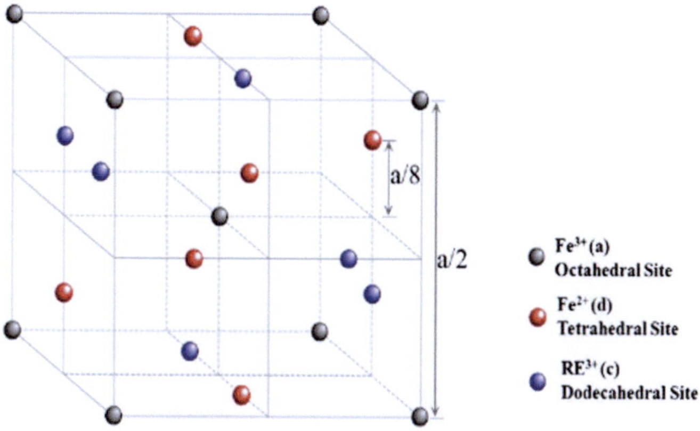

Fig. 1.4 Schematic representation of a garnet crystal phase displaying just the locations of cations. Reproduced with permission from [42]

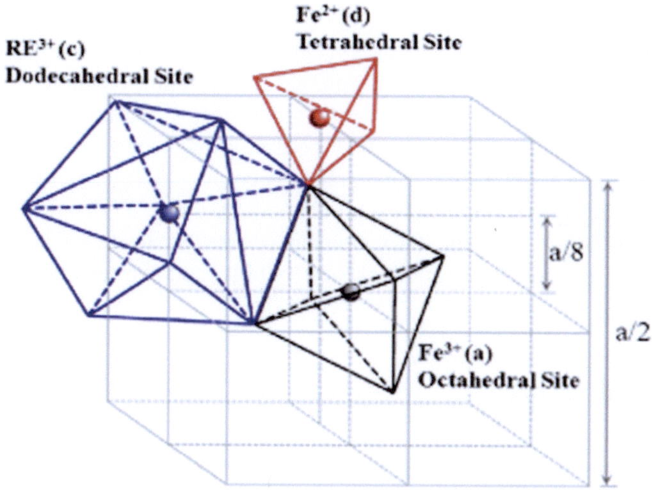

Fig. 1.5 An octant of a garnet crystal phase. Reproduced with permission from [42]

and an electrical system. Orthoferrites are crystallized in a deformed perovskite with an orthorhombic crystal phase, as seen in Fig. 1.6, and exhibit modest ferromagnetism. The mild ferromagnetism arises from little swinging in the configuration involving two antiferromagnetically connected lattices. Antisymmetric exchange process occurs among the vector cross product of neighboring spins rather than a conventional scalar product. Orthoferrites demonstrate antiferromagnetism despite this interaction. They can be engineered for intriguing use in the investigation of laser-induced oscillations and serve as sensors for light [44, 45].

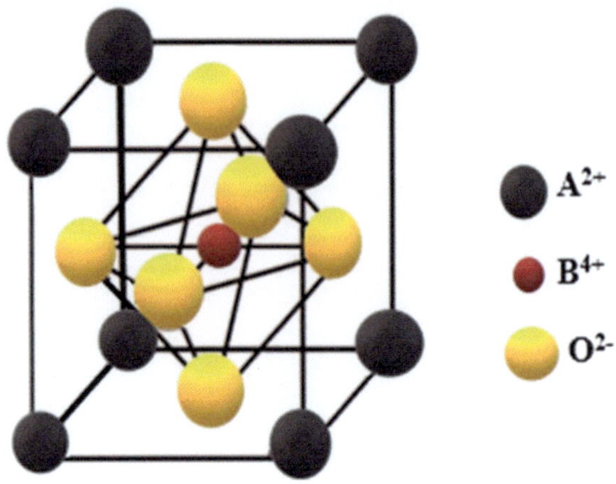

Fig. 1.6 Structure of orthoferrites. Reproduced with permission from [45]

1.1.5 Soft Ferrites

Ferrimagnetic substances that lose their magnetism post-magnetization are referred to as soft ferrites. Soft ferrites have minimal transpiration of energy throughout magnetization, a compact hysteresis cycle, and a steeply ascending magnetization curve (Fig. 1.7a). These substances may be easily magnetized and demagnetized and may be produced by warming and gradually cooling. Soft ferrites are insulating ceramics characterized by the MFe_2O_4 cubic phase, where M denotes a transition metal such as Zn, Ni, and Mn. They are characterized by low eddy current losses, low retentivity and coercivity, and high susceptibility and permeability [44, 46].

1.1.6 Hard Ferrites

Hard ferrites maintain magnetism post-magnetization, so they are designated as long-lasting magnetic substances. Hard ferrites are ferrimagnetic substances distinguished by a substantial hysteresis loop, a progressively ascending magnetization curve, and significant loss of energy throughout the magnetization cycle (Fig. 1.7b). Hard ferrites are challenging to magnetize and demagnetize and may be produced by overheating followed by swift cooling. Iron plus Ba or Sr oxides amalgamate to produce hard ferrites. Moreover, the configuration of hard ferrite is hexagonal. Hard ferrites exhibit substantial retentivity and coercivity, elevated eddy current damage, reduced susceptibility and permeability, and high saturation flux density. Hard ferrites are thus employed in speakers, direct current magnets, long-lasting magnets, and similar applications [8, 46].

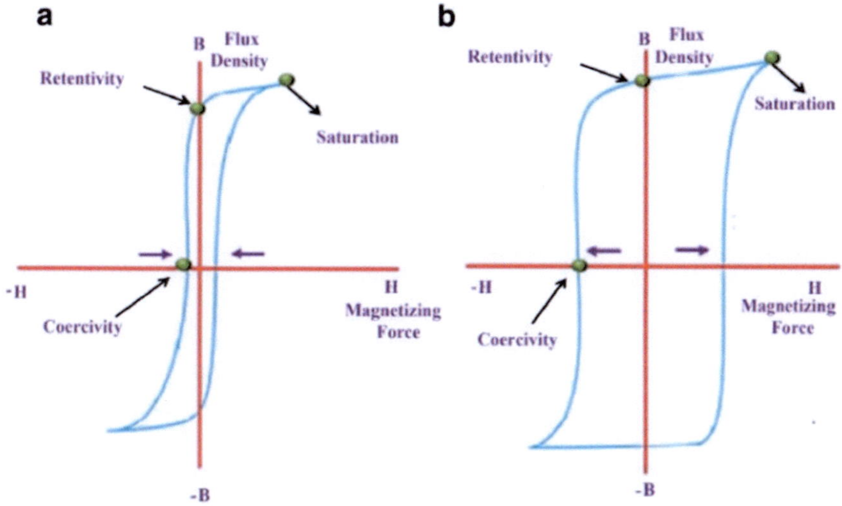

Fig. 1.7 Hysteresis loop of **a** soft and **b** hard ferrites. Reproduced with permission from [44]

1.2 Properties of Ferrites

FNMs exhibit two primary characteristics: they are either insulating or semiconducting, while also being magnetically permeable. When a magnetic field (MF) is applied, the magnetization of ferromagnetic material increases from zero to saturation. FNMs have elevated permeability, substantial saturation magnetization, and dielectric characteristics. FNMs, while capable of transmitting EM radiation at specific frequencies, are often non-conductive in terms of electricity. Chemical makeup, replacements or doping, fabrication techniques, and synthesis circumstances dictate the charge transportation characteristics of FNMs. The dielectric characteristics of FNMs remain consistently constant at elevated frequencies, and their high resistivity results in minimal absorption inefficiencies and other advantageous electrical qualities [47, 48].

1.2.1 Magnetic Properties of FNMs

A variety of variables, including morphology, dimensions, chemical composition, crystalline size, and interparticle connections, affect the magnetic properties of NMs. Magnetic NMs are prevalent in the natural world, present in flora, fauna, and the human brain. The magnetic characteristics of bulk materials, such as the saturation magnetization, Curie temperature, and coercivity, are exclusively determined by crystallographic structure and chemical composition; but, in NMs, identical characteristics are influenced by surface variations and the limited dimension of the particles.

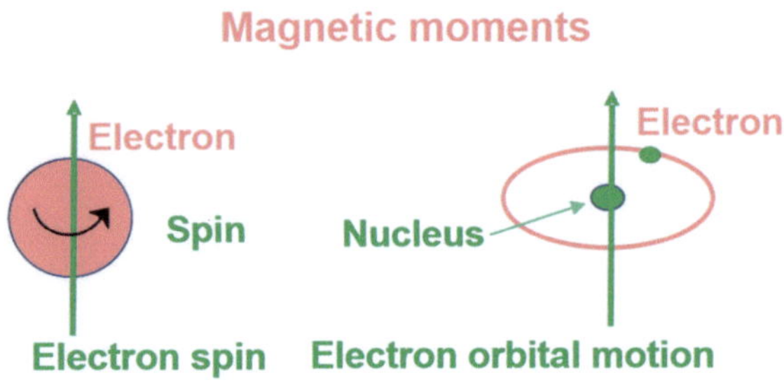

Fig. 1.8 Magnetic moments resulting from e^- spin (left) and e^- orbital motion (right). Reproduced with permission from [48]

Apparent magnetism originates from magnetic moments linked to distinct e^-s. The origins of these magnetic moments are twofold: e^-s spin and orbital motion. e^-spin refers to the rotation of e^-s about their axes, whereas e^-s orbital motion denotes the motion of e^-s in relation to the nucleus (Fig. 1.8). The overall magnetic moment is the aggregate of the individual moments of all e^-s inside the system [49–51].

The magnetic characteristics of a substance are dictated by the reactions of dipoles to an external MF and may be categorized into five types: diamagnetic, ferrimagnetic, antiferromagnetic, ferromagnetic, and paramagnetic, as seen in Fig. 1.9. Diamagnetism, often a feeble kind of magnetism, manifests solely in the presence of an external MF. A change in the e^-orbital velocity creates a counteracting field to the applied MF, resulting in a magnetic moment. The partial cancellation of magnetic moments due to e^- spin and orbital mobility leads to paramagnetism. Ferromagnetism manifests in substances with persistent magnetic moments, existing independently of an externally supplied MF. Common ferromagnetic materials comprise transition metals, particularly those of iron, cobalt, and nickel. The existence of unpaired e^- and incomplete e^- spins results in the elevated magnetic susceptibility of ferromagnetic materials. In antiferromagnetic minerals such as hematite, the spins are oriented in opposite directions, so inhibiting the formation of a net magnetic moment. Ferrimagnetic substances demonstrate a reduced saturation magnetization compared to ferromagnets due to the predominance of spin amplitude in a single direction, leading to net persistent magnetization. Certain ferrites, such as magnetite, demonstrate ferrimagnetism [48, 51].

1.3 Applications of FNMs

FNMs have garnered interest in prominent study domains of biological sensors and sensors, including electrochemical, piezoelectric, optical, and MF (Fig. 1.10). FNMs, including sensors and biosensors, are predominantly employed for the detection and sensing of different analytes at small amounts due to their capacity to orient in accordance with external MF. They may be effectively immobilized on a transducer surface or disseminated within a solution of target analytes for identification purposes [52]. FNMs are well-established chemicals in analytical biochemistry,

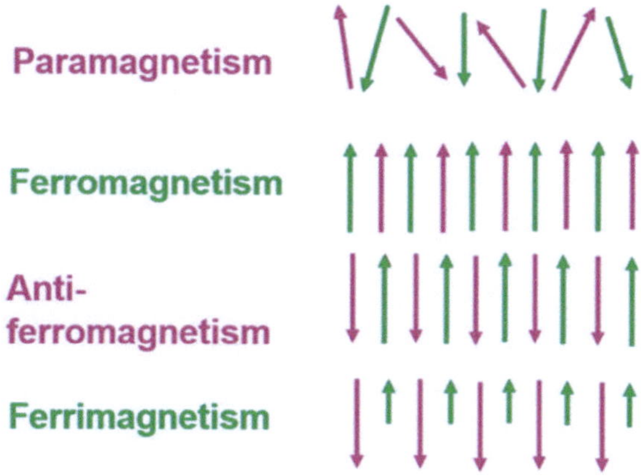

Fig. 1.9 Various types of magnetism. Reproduced with permission from [48]

biotechnology, and pharmaceutical research, owing to their biocompatibility, robust superparamagnetic characteristics, low environmental impact, and facile synthesis. FNMs and their composites with various metal oxides/polymers/2D materials are intensively studied for sensing purposes. Modified FNMs sensing has been developed to leverage two energy-level systems and alter charge transporters for enhanced sensing performance. In the field of ecology, FNMs function as efficient adsorption agents for the elimination of contaminants such as dyes, pesticides, antimicrobial agents, and aromatic substances from drainage. Their dye reduction characteristics facilitate the effective breakdown of organic contaminants under light irradiation. Moreover, their superparamagnetic characteristics provide straightforward segregation and recoveries using external MF, rendering them recyclable and durable throughout numerous cycles [53, 54].

FNMs, especially those composed of $CoFe_2O_4$, are essential in Li and Na sodium batteries and conversion technologies (hydrogen evolution reaction). They are extensively utilized in rechargeable batteries and supercapacitors because of their superior electrochemical characteristics, rigidity, and capability for effective energy storage and discharge. These materials facilitate progress in renewable energy technology, endorsing environmentally friendly energy alternatives [55].

FNMs are predominantly utilized as inductive elements in a wide array of electronic devices, including low-noise amplifiers, filters, and impedance matching networks. Their latest usage as inductors adheres, along with other trends, to the overarching movement of downsizing and incorporating as ferrite multilayers for passive functioning electronic devices. Multilayered technologies have emerged as a crucial method for the mass manufacture of integrated devices, allowing a high level of integration density. Multilayer capacitors entered the market a long time ago, whereas inductors originated in the 1980s. The fundamental elements required to generate inductance are a highly permeable magnetite and a metallic coil [56, 57].

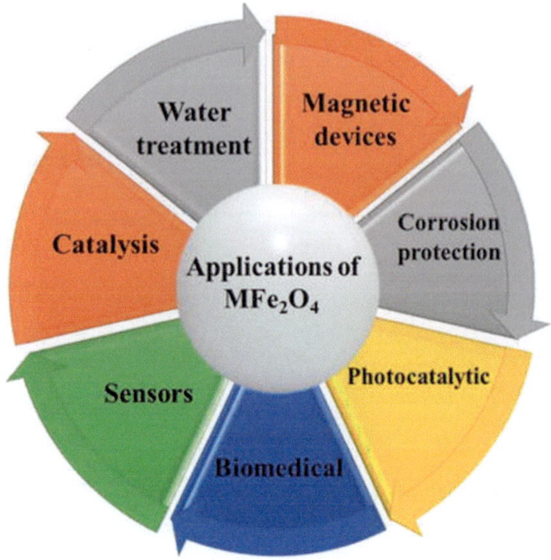

Fig. 1.10 Applications of FNMs. Reproduced with permission from [55]

FNMs characterized by diminutive particle dimensions, readily accessible surface-active sites, elevated capacities for adsorption, facile separation, and substantial surface area to volume ratio render them appropriate materials for the adsorption of contaminants from contaminated water.

Phenols, insecticides, cationic and anionic dyes, and a wide array of aromatic substances are common contaminants in contaminated water [58]. A variety of methods have been utilized to eliminate these contaminants, including reverse osmosis, advanced oxidation, anaerobic treatment, filtering, and adsorption approaches. Nonetheless, several issues hinder their deployment, including challenges in adsorbent recovery, inadequate pollutant removal, and high costs; in certain instances, waste sludge necessitates a substantial disposal area and demands appropriate design. Considering the previously discussed limitations, FNMs emerge as the most appropriate catalyst for tackling water quality and purification challenges [59–61].

FNMs, particularly Fe_3O_4 NPs, are found in several living creatures and have different uses. Fe_3O_4 can be synthesized in the laboratory using established techniques; nevertheless, biogenic FNMs exhibit outstanding characteristics compared to their synthetic counterparts: they have a specific size range and aspect ratio, excessive chemical purity, near-perfect crystallographic structure, and occasionally exhibit unique crystallographic morphological characteristics [62]. The extracellular synthesis of nanometer-sized Fe_3O_4 NPs by several bacterial species has been documented. In some instances, biogenic particles possess a lipid covering that confers stability and enhances biocompatibility [63]. Fe_3O_4 NPs are applicable in several domains, including the transformation, identification, isolation, and examination of cells, as well as the isolation of physiologically active chemicals [64, 65].

References

1. Snelling, E.C.: Soft ferrites: properties and applications (1969)
2. Owens, C.D.: A survey of the properties and applications of ferrites below microwave frequencies. Proc. IRE **44**(10), 1234–1248 (1956)
3. Wijn, H.: Some remarks on the history of ferrite research in Europe. In: Proceedings of the International Conference on Ferrites (1970)
4. Bharathi, K.K., et al.: Effect of annealing on the microstructure of NiFe1. 925Dy0. 075O4 thin films. Thin Solid Films **520**(6), 1794–1798 (2012)
5. Kharisov, B.I., Dias, H.R., Kharissova, O.V.: Mini-review: Ferrite nanoparticles in the catalysis. Arab. J. Chem. **12**(7), 1234–1246 (2019)
6. Sugimoto, M.: The past, present, and future of ferrites. J. Am. Ceram. Soc. **82**(2), 269–280 (1999)
7. Smit, J., Wijn, H.: Physical properties of ferrimagnetic oxides in relation to their technical applications. Philips Techn. Lib. Eindhoven (1959)
8. Abidin, M.Z.U., et al.: A comprehensive review on the synthesis of ferrite nanomaterials via bottom-up and top-down approaches advantages, disadvantages, characterizations and computational insights. Coord. Chem. Rev. **520**, 216158 (2024)
9. Dasgupta, S., et al.: Toward on-and-off magnetism: reversible electrochemistry to control magnetic phase transitions in spinel ferrites. Adv. Func. Mater. **26**(41), 7507–7515 (2016)
10. Zhao, Q., et al.: Spinels: controlled preparation, oxygen reduction/evolution reaction application, and beyond. Chem. Rev. **117**(15), 10121–10211 (2017)
11. Li, Y., et al.: Structural disorder controlled oxygen vacancy and photocatalytic activity of spinel-type minerals: a case study of ZnFe2O4. Chem. Geol. **504**, 276–287 (2019)
12. Nawaz, M., et al.: Photo-Fenton reaction for the degradation of sulfamethoxazole using a multi-walled carbon nanotube-NiFe2O4 composite. Chem. Eng. J. **382**, 123053 (2020)
13. Kefeni, K.K., et al.: Synthesis and application of hematite nanoparticles for acid mine drainage treatment. J. Environ. Chem. Eng. **6**(2), 1865–1874 (2018)
14. Usman, M., et al.: Magnetite and green rust: synthesis, properties, and environmental applications of mixed-valent iron minerals. Chem. Rev. **118**(7), 3251–3304 (2018)
15. Nadimi, M., et al.: Photodegradation of methylene blue by a ternary magnetic TiO2/Fe3O4/graphene oxide nanocomposite under visible light. Mater. Chem. Phys. **225**, 464–474 (2019)
16. Qin, H., et al.: Spinel ferrites (MFe2O4): synthesis, improvement and catalytic application in environment and energy field. Adv. Coll. Interface. Sci. **294**, 102486 (2021)
17. Mohammed, J., et al.: Lightweight SrM/CCTO/rGO nanocomposites for optoelectronics and Ku band microwave absorption. J. Mater. Sci. Mater. Electron. **30**(4), 4026–4040 (2019)
18. El Ata, A.A., Attia, S.: Dielectric dispersion of Y-type hexaferrites at low frequencies. J. Magn. Magn. Mater. **257**(2–3), 165–174 (2003)
19. Pullar, R.C., et al.: Magnetic wood-based biomorphic Sr3Co2Fe24O41 Z-type hexaferrite ecoceramics made from cork templates. Mater. Des. **82**, 297–303 (2015)
20. Al-Garalleh, G.A., et al.: Structural and magnetic properties of RE-Al substituted nanocrystalline hexaferrites (Sr1− xRExAl2Fe10O19). Mater. Res. Express **7**(2), 026103 (2020)
21. Mohammed, I., Mohammed, J., Srivastava, A.K.: Recent progress in hexagonal ferrites based composites for microwave absorption. Cryst. Res. Technol. **58**(3), 2200200 (2023)
22. Bielan, Z., et al.: Application of spinel and hexagonal ferrites in heterogeneous photocatalysis. Appl. Sci. **11**(21), 10160 (2021)
23. Baykal, A., et al.: Pb substituted Ba, Sr-hexaferrite nanoparticles as high quality microwave absorbers. Ceram. Int. **43**(16), 14023–14030 (2017)
24. Trukhanov, A., et al.: Control of electromagnetic properties in substituted M-type hexagonal ferrites. J. Alloy. Compd. **754**, 247–256 (2018)
25. Yang, Y., et al.: A study on structural, spectral, and magnetic properties of Pr–Bi co-doped M-type barium–strontium hexaferrites via the solid-state reaction method. Appl. Phys. A **124**(12), 854 (2018)

26. Kaur, T., et al.: Enhancement in physical properties of barium hexaferrite with substitution. J. Mater. Res. **30**(18), 2753–2762 (2015)
27. Alam, R.S., et al.: Structural, magnetic and microwave absorption properties of doped Ba-hexaferrite nanoparticles synthesized by co-precipitation method. J. Magn. Magn. Mater. **381**, 1–9 (2015)
28. Chang, S., Kangning, S., Pengfei, C.: Microwave absorption properties of Ce-substituted M-type barium ferrite. J. Magn. Magn. Mater. **324**(5), 802–805 (2012)
29. Yasmin, N., et al.: Influence of samarium substitution on the structural and magnetic properties of M-type hexagonal ferrites. J. Magn. Magn. Mater. **446**, 276–281 (2018)
30. Narang, S.B., Pubby, K., Singh, C.: Thickness and composition tailoring of K-and Ka-band microwave absorption of BaCo x Ti x Fe (12–2 x) O19 ferrites. J. Electron. Mater. **46**(2), 718–728 (2017)
31. Li, Z., Gao, F.: Chemical bond and hardness of M-, W-type hexagonal barium ferrites. Can. J. Chem. **89**(5), 573–576 (2011)
32. Hsiang, H.-I., Yao, R.-Q.: Hexagonal ferrite powder synthesis using chemical coprecipitation. Mater. Chem. Phys. **104**(1), 1–4 (2007)
33. Smit, J., Wijn, H.: Ferrites, Philips technical library. Eindhoven, The Netherlands **278** (1959)
34. Jones, G., et al.: Domain structure of the single crystal hexagonal ferrite, Co/sub 2/X. IEEE Trans. Magn. **26**(5), 2804–2806 (2002)
35. Cao, H., et al.: High pressure floating zone growth and structural properties of ferrimagnetic quantum paraelectric BaFe12O19. Appl. Mater. **3**(6) (2015)
36. Smit, J., Wijn, H.: Ferrites Philips Techn. Lib. Eindhoven. The Netherlands **151**, 157–158 (1959)
37. Castelliz, L., Kim, K., Boucher, P.: Preparation, stability range and high frequency permeability of some ferroxplana compounds. J. Can. Ceram. Soc **38**, 57 (1969)
38. Pullar, R.C.: Hexagonal ferrites: a review of the synthesis, properties and applications of hexaferrite ceramics. Prog. Mater. Sci. **57**(7), 1191–1334 (2012)
39. Modi, K., et al.: On the relationship between structural-optical parameters of Y3− xFe5+ xO12 garnet ferrites and the oxide additivity rule. J. Supercond. Novel Magn. **29**(7), 1931–1936 (2016)
40. Akhtar, M.N., et al.: Impacts of Gd–Ce on the structural, morphological and magnetic properties of garnet nanocrystalline ferrites synthesized via sol–gel route. J. Alloy. Compd. **660**, 486–495 (2016)
41. Geller, S., Gilleo, M.: The crystal structure and ferrimagnetism of yttrium-iron garnet, Y3Fe2 (FeO4) 3. J. Phys. Chem. Solids **3**(1–2), 30–36 (1957)
42. Jadhav, V.V., et al.: Properties of ferrites. In: Spinel Ferrite Nanostructures for Energy Storage Devices, pp. 35–50. Elsevier (2020)
43. Bozorth, R., Williams, H., Walsh, D.E.: Magnetic properties of some orthoferrites and cyanides at low temperatures. Phys. Rev. **103**(3), 572 (1956)
44. Ahmed, M.M., Abu-Elsaad, N.: Exploring the magnetic behavior of ferrites: from diamagnetism to superparamagnetism (2024). arXiv:2406.10599
45. Sambhudevan, S.: Ferrite-based polymer nanocomposites as shielding materials: a review. Chem. Pap. **75**(8), 3697–3710 (2021)
46. Vedrtnam, A., et al.: A comprehensive study on structure, properties, synthesis and characterization of ferrites. AIMS Mater. Sci. **7**(6), 800–835 (2020)
47. Tolani, S.C., Golhar, A., Rewatkar, K.: A review of morphological, structural behaviour and technological applications of ferrites. In: AIP Conference Proceedings. AIP Publishing LLC (2019)
48. Abdulwahab, K.O., Khan, M.M., Jennings, J.R.: Ferrites and ferrite-based composites for energy conversion and storage applications. Crit. Rev. Solid State Mater. Sci. **49**(5), 807–855 (2024)
49. Suciu, M., et al.: Applications of superparamagnetic iron oxide nanoparticles in drug and therapeutic delivery, and biotechnological advancements. Beilstein J. Nanotechnol. **11**(1), 1092–1109 (2020)

50. Abdulwahab, K.O., et al.: A one-pot synthesis of monodispersed iron cobalt oxide and iron manganese oxide nanoparticles from bimetallic pivalate clusters. Chem. Mater. **26**(2), 999–1013 (2014)
51. Gore, S.K., Jadhav, S.S.: Basics of ferrites. In: Spinel Ferrite Nanostructures for Energy Storage Devices, pp. 1–11. Elsevier (2020)
52. Srinivasan, S.Y., et al.: Applications of cobalt ferrite nanoparticles in biomedical nanotechnology. Nanomedicine **13**(10), 1221–1238 (2018)
53. Shobana, M.: Nanoferrites in biosensors–a review. Mater. Sci. Eng. B **272**, 115344 (2021)
54. Ranga, R., et al.: Ferrite application as an electrochemical sensor: a review. Mater Charact **178**, 111269 (2021)
55. Li, H., et al.: Magnetic spinel ferrite nanomaterials: fabrication and application (2025)
56. Yang, C., et al.: Fully integrated ferrite-based inductors for RF ICs. Sens. Actuators A **130**, 365–370 (2006)
57. Valenzuela, R.: Novel applications of ferrites. Phys. Res. Int. **2012**(1), 591839 (2012)
58. Zahid, M., et al.: Metal ferrites and their graphene-based nanocomposites: synthesis, characterization, and applications in wastewater treatment. In: Magnetic Nanostructures: Environmental and Agricultural Applications, pp. 181–212. Springer, Berlin (2019)
59. Bhatnagar, A., et al.: A review of the use of red mud as adsorbent for the removal of toxic pollutants from water and wastewater. Environ. Technol. **32**(3), 231–249 (2011)
60. El-Moselhy, M.M., Sengupta, A.K., Smith, R.: Carminic acid modified anion exchanger for the removal and preconcentration of Mo (VI) from wastewater. J. Hazard. Mater. **185**(1), 442–446 (2011)
61. Dhir, B.: Potential of biological materials for removing heavy metals from wastewater. Environ. Sci. Pollut. Res. **21**(3), 1614–1627 (2014)
62. Lowenstam, H.A.: Magnetite in denticle capping in recent chitons (Polyplacophora). Geol. Soc. Am. Bull. **73**(4), 435–438 (1962)
63. Zhang, C., et al.: Formation of single-domain magnetite by a thermophilic bacterium. Am. Miner. **83**(11), 1409–1418 (1998)
64. Šafařík, I., Šafaříková, M.: Use of magnetic techniques for the isolation of cells. J. Chromatogr. B Biomed. Sci. Appl. **722**(1–2), 33–53 (1999)
65. Sode, K., et al.: Application of bacterial magnetic particles for highly selective mRNA recovery system. Biotechnol. Tech. **7**(9), 688–694 (1993)

Chapter 2
Synthesis Protocol

2.1 Bottom-Up Approaches

2.1.1 Co-precipitation

Coprecipitation is widely adopted simple, cost-effective method of synthesizing spinel ferrite nanoparticles (NPs). This technique was first introduced by Massart, since it has been extensively used for nanomaterial synthesis. In this method, metal ions are precipitated from a solution using an appropriate base (NaOH). This approach allows control oversize and distribution by controlling the nucleation and growth rates. It is particularly valued for its ability to prepared ferrite NPs with high purity, uniform morphology, and controlled size at relatively low temperature. It is an energy efficient method used for various applications highlighting its potential for research and industrial applications. Synthesis parameters like pH, temperature, and precursor can be easily adjusted for tailored material characteristics. The arrangement of metal ions in ferrites with inverse spinel structure follows a specific pattern, M^{2+} ions (Ni^{2+}, Co^{2+}, or Zn^{2+}) occupy the octahedral sites, while the Fe^{3+} ions occupies octahedral and tetrahedral sites within the crystal lattice. This cation distribution significantly influences the properties of material (Fig. 2.1) [1–3]. This method has been extensively utilized since the early 1980s [3] for fabricating magnetic nanomaterials with high crystallinity and desirable grain sizes. In this method, divalent metal salts (e.g. Mg^{2+}, Co^{2+}) and trivalent metal salts (e.g., Fe^{3+}) such as chlorides, nitrates, sulfates, or sulfites are mixed in a 1:2 molar ratio (M^{2+}: Fe^{3+}). The precipitating agent (NaOH, NH_4OH) was added to the solution to induce metal hydroxide precipitates formation. The resulting precipitates were acquired by filtration or centrifugation and washed with distilled water or with other solvents to remove contaminants. The material was then dried at 150 and 300 °C. The reaction commonly takes place in basic medium, and the pH of solution was carefully monitored to ensure uniform nucleation and particle growth. This influences the final particle size, distribution, and structural properties. A key advantage of this method is that it does not require

M. Ikram et al., *Ferrite Nanomaterials for Wastewater Treatment*, Engineering Materials, https://doi.org/10.1007/978-3-032-09141-3_2

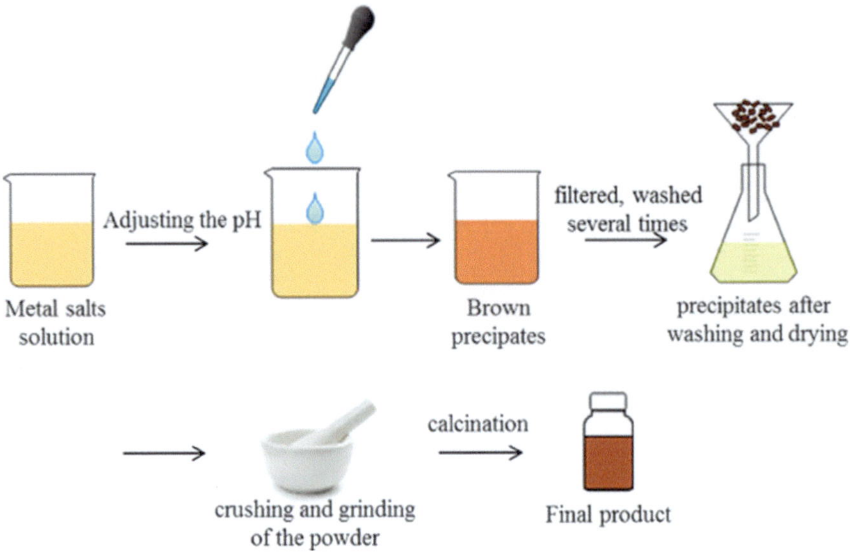

Fig. 2.1 Methodological sequence for the synthesis of spinel ferrite using the coprecipitation method. Reproduced with permission from [7]

organic fuels like citric acid and produces fine, uniformly distributed nanoparticles directly from the solution [4–7]. Table 2.1 indicates the co-precipitation synthesis, morphology, and crystal structure of FNMs.

For instance, we take an example of $MnFe_2O_4$ nanoparticle synthesis using a coprecipitation route. The NPs formation can be attributed to the chemical reaction as described by Eqs. (2.1)–(2.7). With NaOH in addition to the metal salts, the metal hydroxide was formed according to the Eqs. (2.4)–(2.5) [8];

$$MnSO_4.H_2O \rightarrow Mn^{2+}_{4(aq)} + SO^{(2-)}_{4(aq)} + H_2O_{(aq)} \tag{2.1}$$

$$2FeSO_4.7H_2O(s) \rightarrow 2Fe^{2+}_{(aq)} + 2SO^{(2-)}_{4(aq)} + 7H_2O_{(aq)} \tag{2.2}$$

$$4NaOH_{(s)} \leftarrow 4Na^{+}_{(aq)} + 4OH^{-}_{(aq)} \tag{2.3}$$

$$Mn^{2+}_{(aq)} + 2OH^{-}_{(aq)} + xH_2O_{(aq)} \rightarrow Mn(OH)_2.xH_2O_{(s)} \tag{2.4}$$

$$2Fe^{2+}_{(aq)} + 4OH^{-}_{(aq)} + xH_2O \rightarrow 2Fe(OH)_2.xH_2O_{(s)} \tag{2.5}$$

$$Mn(OH)_2.xH_2O_{(s)} + 2Fe(OH)_2.xH_2O_{(s)} \rightarrow Mn(OH)_{2(s)} + 2Fe(OH)_{2(s)} + xH_2O_{(g)} \tag{2.6}$$

Table 2.1 Synthesis of ferrites by coprecipitation method

Sr. No.	Material	Precursor	Size	Crystal structure	Morphology	References
1	$CoFe_2O_4$	$FeCl_3{\cdot}6H_2O$, $CoCl_2{\cdot}6H_2O$	Particle size = 2–14 nm	Cubic	NPs	[9]
2	$ZnFe_2O_4$	$ZnCl_2$, $FeCl_3$	Crystallite size = 5–11 nm	Cubic	NPs	[10]
3	$NiFe_2O_4$	$Ni(NO_3)_2$, $Fe(NO_3)_3$	Crystallite size = 8–20 nm, Particle size = 10–20 nm	Cubic	NPs	[11]
4	$NiFe_2O_4$	$NiCl_2$, $FeCl_3$	Particle size = 8–28 nm	Cubic	NPs	[12]
5	$MgFe_2O_4$	$MgCl_2.6H_2O$, $FeCl_3.\ 6H_2O$	Grain size = 2.1–10.7 nm	–	NPs	[13]
6	$CuFe_2O_4$	$Cu(NO_3)_2.6H_2O$, $FeCl_3{\cdot}6H_2O$	Grain size = 10–90 nm	Cubic	NPs	[14]
7	$SrFe_2O_4$	$Sr(NO_3)_2$, Fe $(NO_3)0.9H_2O$	Particle size = 112–120 nm	Cubic	NPs	[15]
8	$SnFe_2O_4$	$FeCl_3{\cdot}6H_2O$, $SnCl_2{\cdot}2H_2O$	Grain size = 38.9–89 nm	Cubic	Cubic and polyhedral	[16]
9	$BiFeO_3$	$Bi(NO_3)_3{\cdot}5H_2O$, $Fe(NO_3)_3{\cdot}9H_2O$	Particle size = 45 nm	Rhombohedral	NPs	[17]
10	$CrFe_2O_4$	Iron sand, $CrCl_3.6H_2O$	Particle size = 27.6–34.2 nm	Cubic	NPs	[18]

$$Mn(OH)_{2(s)} + 2Fe(OH)_{2(s)} \rightarrow MnFe_2O_{4(s)} + 3H_2O_{(g)} \tag{2.7}$$

2.1.2 Sol–Gel

The sol–gel technique is a widely employed versatile method for materials synthesis like metal oxides, nitrides, and carbides (Fig. 2.2). The process begins by preparing a "sol," a homogeneous solution by dissolving metal alkoxides or metal salts in either an alcohol or a water-alcohol mixture. The technique involves hydrolysis and condensation reactions. During hydrolysis metal alkoxide (M-OR) reacts with water forms metal hydroxide (M-OH) species within a colloidal solution. These -OH groups then participates in condensation reactions, resulting in the formation of M-O-M linkages and releasing byproducts (water or alcohol) [19, 20]. This creates a

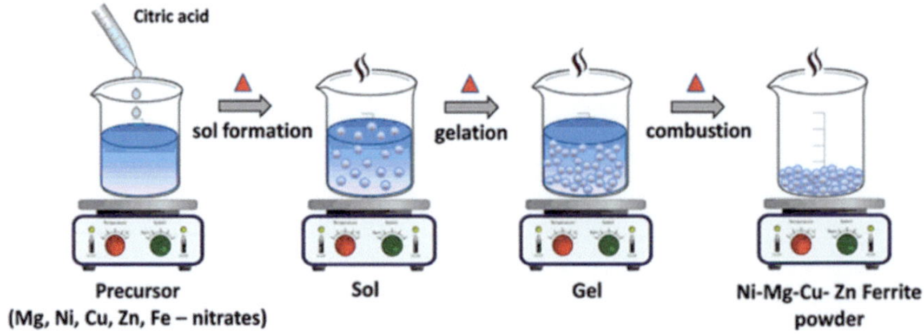

Fig. 2.2 Schematic illustration of ferrite gel preparation by sol–gel. Reproduced with permission from [29]

3D network, transforming the sol into a gel forming a semi-solid matrix consists of liquid and solid components [21, 22]. Factors including solution pH, precursor type, temperature, and solvent concentration significantly influence the reactions. At lower pH the positively charged transition states are stabilized by -OR groups, slow down hydrolysis and resulting in the formation of larger, less branched particles. In contrast, at high pH the negatively charged intermediates promote condensation, forming smaller particles. As particle size decreases the Van der Waals forces dominate and resulted in aggregation or cluster formation. Hence, precise control of parameters are essential for gel to stabilize and ensure a large surface area [23, 24].

After gelation, the remaining solvents were removed through drying. Numerous drying techniques were employed which modifies the gel, forming a dense xerogel. Supercritical drying maintains the original structure results in an aerogel, while freeze drying removes solvent forming a cryogel. After drying the calcination is done to remove residual organic content. This step is carried out at temperatures of 300–800 °C in oxygen rich environment. Calcination is crucial in defining crystallinity, particle size, and overall density of the product. This results in fine uniformly distributed and pure oxide powder [25, 26].

Citrate-based sol–gel method is the modification of the conventional sol–gel technique. In this method, metal nitrates and citric acid are mixed with 1:1 molar ratio, and the solution pH is adjusted to ~7. The reaction mixture is then heated at ~80 °C, formatting a viscous gel. Upon ignition, this gel underwent a combustion reaction and yields a loose powder and then calcined to obtain ferrite NPs [27]. This method yields fine particles of ~10 nm at low temperatures. The sol–gel method has been employed to prepare an array range of spinel ferrites [28]. The structural and electrical characteristics of ferrites are influenced by synthesis parameters like sintering temperature, sintering duration, and the presence of dopants. In cases where alkoxides are unavailable, salts such as acetates can serve as alternative precursors [19, 23]. Despite its advantages the sol–gel process has some limitations including the high cost and contamination from reaction byproducts. However, its versatility, precision, and eco-friendly nature makes the sol–gel method to be one of the most promising approaches for the controlled synthesis of ferrite nanomaterials. Table 2.2 exhibits the sol–gel synthesis, morphology, and crystal structure of FNMs.

2.1.3 *Solvothermal*

The solvothermal method is an efficient and versatile approach for FNMs synthesis offer control size distribution, morphology and crystalline structure (Fig. 2.3). In this method, metal precursors (metal alkoxides or salts) were dissolved into solvents with the help of mineralizers such as metal hydroxide. The resulting solution was then sealed in an autoclave and dried at an elevated temperature (100–1000 °C) and pressure of (1–1000 atm) to enhance interaction and facilitate nucleation and growth. After that the resulting product was centrifuged, washed with deionized water and subjected to drying process to obtain powder form. The method referred as hydrothermal because we water as a solvent [39, 40].

This method offers the advantage of tuning material properties by controlling parameters like temperature, reaction time, solvent, surfactants, and precursor ratio. This method has been employed to synthesize an array of ferrites and their composites [41–43]. The NPs formed are monodisperse, low in agglomeration, have high saturation, and possess substantial surface area. These properties are highly desirable for industrial and biomedical applications [43, 44].

It is a simplicity, cost-effective, and ecofriendly route enables the synthesis of materials with excellent crystallinity, high quality with less impurities without complex instrumentation [43, 44]. However, the main drawback is the relatively long reaction time [22]. Apart from its limitations, the solvothermal process remains one of the most promising and widely used techniques for producing FNMs with tailored physiochemical properties making it promising approach for advanced technological applications. Table 2.3 represents the solvothermal synthesis, morphology, and crystal structure of FNMs.

2.1.4 *Spray Pyrolysis*

Spray pyrolysis is a widely employed, cost-effective, and adaptable deposition method for producing ceramic coatings, powders, and dense or porous oxide layers. It does not require expensive substrates unlike other deposition method, making it promising approach for fabricating multilayered structures and powder-based materials [53, 54].

The process involves the preparation of a precursor (metal salts such as chlorides, nitrates, acetates, or acetylacetonates) solution. The salt solubility, droplet spreading, and solvent volatility significantly influence the solution behavior [55, 56]. The choice of solvent influences the precursor, flow rate, and the temperature at which the precursor decomposes into oxides. The solution was then atomized

Table 2.2 Synthesis of ferrites by sol–gel method

Sr. no.	Material	Precursor	Size	Crystal structure	Morphology	References
1	$CoFe_2O_4$	$Fe(NO_3)_3{\cdot}9H_2O$, $Co(NO_3)_2.6H_2O$	Particle size = 7–28 nm	Cubic	NPs	[30]
2	$MgFe_2O_4$	$Fe(NO_3)_3{\cdot}9H_2O$, $Mg(NO_3)_2{\cdot}6H_2O$	Grain size = 162 nm	Cubic	NPs	[31]
3	$MnFe_2O_4$	$Mn(NO_3)_2$, $Fe(NO_3)_3{\cdot}9H_2O$		Rhombohedral	NPs	[32]
4	$BaFe_2O_4$	$Ba(NO_3)_2$, $Fe(NO_3)_3{\cdot}9H_2O$	Particle size = 70 nm	Orthorhombic	NPs	[33]
5	$ZnFe_2O_4$	$Zn\ (NO_3)_2$, $Fe\ (NO_3)_3$	Crystallite size = 11 nm	Cubic	NPs	[34]
6	$NiFe_2O_4$	$Fe(NO_3)_3.9H_2O$, $Ni\ (NO_3)_2.6H_2O$	Crystallite size = 113 nm	Cubic	NPs	[35]
7	$SnFe_2O_4$	$FeCl_3.6H_2O$, $SnCl_2.2H_2O$	Grain size = 87–92 nm	Cubic	NPs	[16]
8	$SrFe_2O_4$	$Fe(NO_3)_3.9H_2O$, $Sr\ (NO_3)_2$	–	Cubic	NPs	[36]
9	$SrFe_2O_4$	$Fe(NO_3)_3{\cdot}9H_2O$, $Sr(NO_3)_2$	Particle size = 30–40 nm	Cubic	NPs	[37]
10	$CuFe_2O_4$	$Cu(NO_3)_2.3H_2O$, $Fe(NO_3)_3.9H_2O$	Crystallite size = 37.34 nm	Tetragonal	NPs	[38]

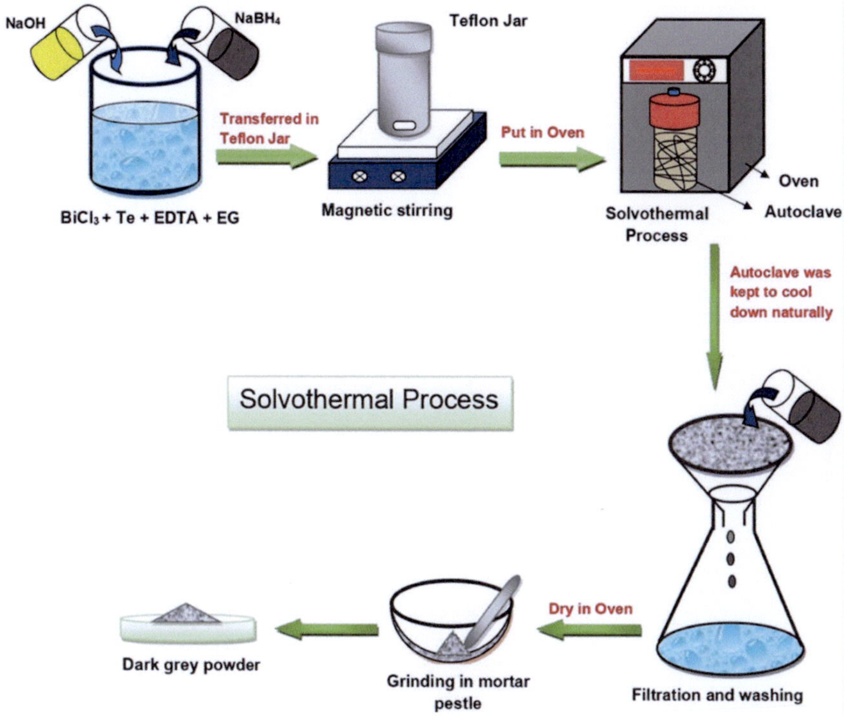

Fig. 2.3 Synthesis of ferrite via solvothermal route. Reproduced with permission from [45]

into fine droplets. The atomization is crucial in determining the quality of the ferrite films. Size, velocity of droplet, and rate of atomization effects how the droplets heated, evaporate, and interact with the substrate. Smaller droplets were achieved at higher pressures that leads to more uniform and finely structured films. The droplet properties are influenced by the viscosity, surface tension, and density of solution [57, 58].

As droplets approach the heated substrate, solvent evaporation begins. This evaporation is influenced by droplet temperature, diffusion of solvent vapor into the gas phase, and solute migration toward the droplet center [53, 57, 58]. The stoichiometry and uniformity of the resulting film are highly sensitive to this evaporation stage. Ideally, solvent evaporation should complete just before the droplet impacts the substrate. The mechanism involves multiple steps, including atomization, evaporation, formation of precipitates, pyrolytic decomposition of precipitate, nucleation on substrate, and growth of film layers [58].

Substrate temperature significantly influences synthesis by spray pyrolysis. Lower substrate temperatures result in porous and amorphous ferrite films, in contrast to

Table 2.3 Synthesis of ferrites by the solvothermal method

Sr. no.	Material	Precursor	Size	Crystal structure	Morphology	References
1	$CoFe_2O_4$	$Fe(NO_3)_3 \cdot 9H_2O$, $Co(NO_3)_2 \cdot 6H_2O$	Particle size = 5.5–12 nm	Cubic	NPs	[46]
2	$MgFe_2O_4$	$Mg(NO_3)_2 \cdot 6H_2O$, $Fe(NO_3)_3 \cdot 9H_2O$	Diameter = 370 nm	Cubic	Nanosphere	[47]
3	$MnFe_2O_4$	$Mn(CH_3COO)_2$, $FeCl_3$	Diameter = 150–250 nm	Cubic	Microsphere	[48]
4	$ZnFe_2O_4$	$FeCl_3.6H_2O$, $ZnCl_2$	Diameter = 284–345 nm	Cubic	Colloidal nanoparticle assemblies (CNAs)	[41]
5	$ZnFe_2O_4$	$FeCl_3 \cdot 6H_2O$, $ZnCl_2$	Diameter = 330–560 nm	Cubic	CNAs	[49]
6	$NiFe_2O_4$	$FeCl_3 \cdot 6H_2O$, $NiCl_2 \cdot 6H_2O$	Diameter = 140 nm	Cubic	Nanosphere	[50]
7	$SnFe_2O_4$	$SnCl_2.2H_2O$, $FeCl_3.6H_2O$	Average cube size = 288 nm	Cubic	NPs	[51]
8	$CuFe_2O_4$	$CuCl_2 \cdot 2H_2O$, $FeCl_3 \cdot 6H_2O$	Diameter = 200 nm	Cubic	Nanosphere	[52]

higher temperatures they are highly crystalline. Substrate temperature ensures effective breakdown of the precursor into the desired oxide. The morphology and thickness of ferrite film depend on precursor ratio, size and velocity of droplet, temperature, spray, nozzle to substrate distance, gas flow, and the total volume of solution [59]. Droplets with optimal momentum spread evenly, forming a uniform coating. In contrast, droplets with insufficient velocity may simply rest on the surface, while excessively fast droplets may skip off the substrate, leading to poor film uniformity. Precursor concentration has a strong influence on film structure. Solutions with relatively low concentration generally led to smoother, transparent and compact films, with slower growth rates. High concentration solutions promote film growth, produce roughness and inhomogeneities. The solvent influences evaporation, droplet behavior, and breakdown efficiency. Solvents (acetyl alcohol or acetylacetonate) decompose effectively than water based solutions tailoring properties like resistivity and film texture [54, 60]. Hence, spray pyrolysis is a versatile and efficient route for synthesizing ferrite thin films. Table 2.4 represents the spray pyrolysis synthesis, morphology, and crystal structure of FNMs.

Table 2.4 Synthesis of ferrites by spray pyrolysis method

Sr. no.	Material	Precursor	Size	Crystal structure	Morphology	References
1	$CoFe_2O_4$	$Co(NO_3)_2.6H_2O$, Fe $(NO_3)_3.9H_2O$	Grain size = 8–10 nm	Cubic	Mesoporous grain like morphology	[61]
2	$MgFe_2O_4$	$Mg(NO_3)_2.6H_2O$, $Fe(NO_3)_3.9H_2O$	Diameter = 206–340 nm	Cubic	Nanosphere	[62]
3	$MnFe_2O_4$	$Mn(NO_3)_2.4H_2O$, $FeCl_3.6H_2O$	Particle size = 82 nm		NPs	[63]
4	$MnFe_2O_4$	$Mn(NO_3)_2.6H_2O$, Fe $(NO_3)_3.9H_2O$	Crystallite size = 15 nm	Cubic		[64]
5	$NiFe_2O_4$	$Ni(NO_3)_2{\cdot}6H_2O$, $FeCl_3{\cdot}6H_2O$	Crystallite size = 19.5 nm, particle diamater = 80 nm	Cubic	NPs	[65]
6	$ZnFe_2O_4$	$Fe(NO_3)_3{\cdot}9H_2O$, $Zn(NO_3)_2{\cdot}6H_2O$	Pore diameter = 200–220 nm	Cubic	Macroporous spheres (MPs)	[66]
7	$CuFe_2O_4$	$Fe(NO_3)_3{\cdot}9H_2O$ $Cu(NO_3)_2{\cdot}3H_2O$	Particle diameter = 20–30 nm	Cubic	NPs	[67]
8	$BaFe_{12}O_{19}$	$Ba(NO_3)_2$, $Fe(NO_3)_3{\cdot}9H_2O$	Particle diameter = 72 nm	Hexagonal	NPs	[68]

2.1.5 *Hydrothermal*

The hydrothermal synthesis method is a solution-based bottom-up technique that involves the reaction of solid materials with an aqueous solvent under elevated temperature and pressure (Fig. 2.4). It is performed in an autoclave, where the temperature is increased than the water boiling point and achieves vapor saturation. This method allows the dissolution and recrystallization of materials that are otherwise insoluble under normal conditions. It is referred to as “hydrothermal” because water is used as the primary solvent during the synthesis process [69, 70].

Various metal precursors (metal salts, oxides, and hydroxides) were dissolved in solvent [71]. Common precursors are iron salts ($Fe(NO_3)_3{\cdot}9H_2O$ or $FeCl_3{\cdot}6H_2O$) and metal salts ($M(NO_3)_2$, MSO_4, or MCl_2). The pH of the solution is adjusted to 7–12, depending on reaction need. The solution was transferred to an autoclave and heated at 100 and 400 °C for 12–24 h. After that, the system cool naturally at ambient temperature, and the resulting product is collected by filtration/centrifugation and dried at ~85 °C overnight [72, 73]. The process involves dissolving the metal precursors in the solvent to form a homogeneous solution and sealed in an autoclave with maintained temperature difference promoting ion migration and nucleation and then washing and drying the sample to obtain the final product. The particles synthesized using this method are small, uniform, and exhibit structural symmetry. This technique allows controlled nucleation rates and crystal growth by controlling key parameters like temperature, pressure, pH, precursor mass, and time. Hydrothermal synthesis is used for producing homogeneous ferrites with enhanced chemical compositions and reduced grain sintering or agglomeration [74, 75]. This method consumes low energy,

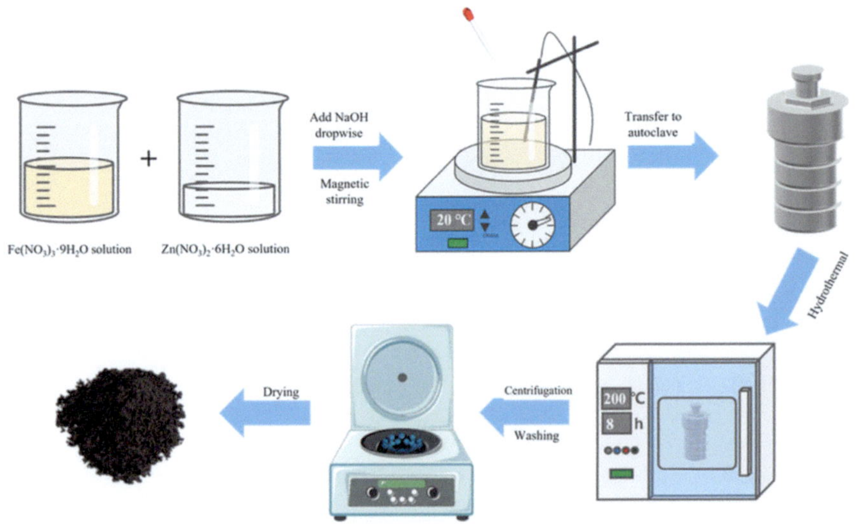

Fig. 2.4 Schematic illustration of $ZnFe_2O_4$ synthesis via hydrothermal synthesis. Reproduced with permission from [77]

is eco-friendly, offers purity, and produces particles with controlled morphology. It offers crystal growth with less particle size and homogeneity. In contrast, the method have some limitations like slow reaction kinetics and high-pressure reactors that restrict its large-scale industrial application [70, 76]. Hydrothermal synthesis, despite these limitations, proves to be an efficient method for advanced ferrite nanomaterials synthesis. Table 2.5 exhibits the hydrothermal synthesis, morphology, and crystal structure of FNMs.

Table 2.5 Synthesis of ferrites by the hydrothermal method

Sr. no.	Material	Precursor	Size	Crystal structure	Morphology	References
1	$CoFe_2O_4$	$FeCl_3{\cdot}6H_2O$, $CoCl_2$	Diameter = 25 nm Length = 120 nm		NRs	[78]
2	$CoFe_2O_4$	$CoCl_2{\cdot}6H_2O$, $FeCl_3{\cdot}6H_2O$	Diameter = 90–180 nm	Cubic	Nanospheres	[79]
3	$MgFe_2O_4$	$Mg(NO_3)_2{\cdot}6H_2O$, $Fe(NO_3)_3.9H_2O$	Crystallit size = 19.5 nm	Cubic	NPs	[80]
4	$MnFe_2O_4$	$FeCl_3$, $MnCl_2$,$4H_2O$	Particle size = 10–20 nm	Cubic	NPs	[81]
5	$ZnFe_2O_4$	$ZnSO_4$, $Zn(NO_3)_2{\cdot}6H_2O$	Particle size = 8–18 nm	Cubic	Nanocrystal	[82]
6	$NiFe_2O_4$	$Ni(NO_3)_3.6H_2O$, $Fe(NO_3)_3.9H_2O$	Particle size = 48–60 nm	Cubic	NPs	[83]
7	$SnFe_2O_4$	$FeCl_3{\cdot}6H_2O$, $SnCl_2{\cdot}2H_2O$	Particle size = 15–50 nm	Cubic	NPs	[84]
8	$BaFe_2O_4$	$Ba(NO_3)_2$, $Fe(NO_3)_3{\cdot}9H_2O$	Crystalllite size = 52 nm Size = 25 nm	Orthorhombic	NSs	[85]
9	$CuFe_2O_4$	$Cu(NO_3)_2{\cdot}3H_2O$, $Fe(NO_3)_3{\cdot}9H_2O$	Grain size = 8–24 nm	Cubic	NPs	[86]
10	$BiFe_2O_4$	BiOCl, $FeCl_3$	Crystallite size = 43.49 nm	Rhombohedral	NPs	[87]

2.2 Top-Down Approaches

2.2.1 Laser Ablation

Laser ablation is a top-down physical method that utilizes a high-energy laser beam to remove material from a solid target (Fig. 2.5). As photons strike the substrate the electrons excite generate heating causing melting and vaporization of surface atoms. The ablated material interacts with medium (liquid) to form NPs with no contamination. Laser sources used in these processes were classified by their emission wavelength (excimer lasers and infrared lasers). Excimer lasers have low power efficiency (0.2–2%) and emitting wavelength of 10–400 nm. F_2, KrF, ArF, CeBr, and XeCl lasers were commonly used. IR (Nd: YAG) lasers with wavelength of 1064 nm and diode lasers with 800–970 nm range were frequently used for ablating polymeric materials (PMMA, PTFE, and PI) [88–90].

Laser ablation is a straightforward technique that facilitates the formation of high purity, monodisperse NPs. This method operates at room temperature and pressure, allows the formation of new phases and offers tailored NSs with enhanced

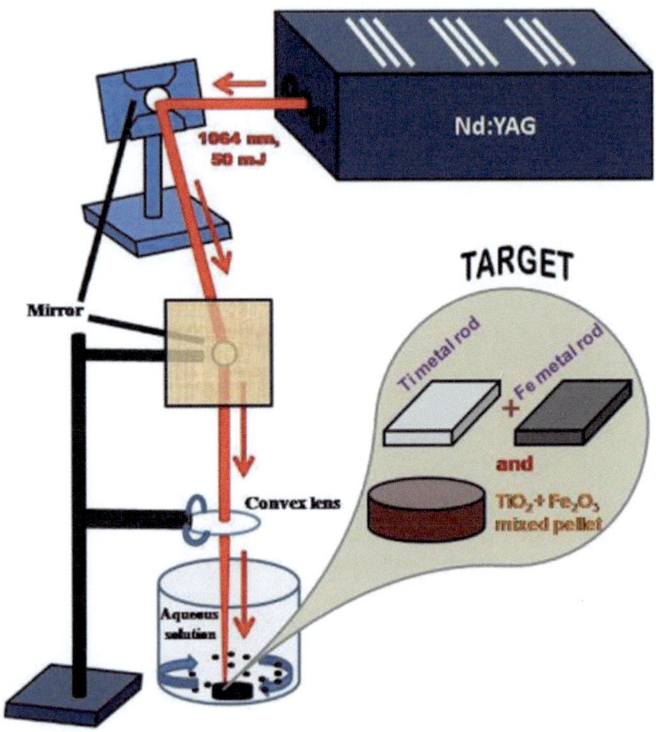

Fig. 2.5 Schematic illustration of $TiFe_2O_4$ synthesis via laser ablation. Reproduced with permission from [100]

optical, magnetic, and catalytic properties. FNMs synthesized by laser ablation exhibit improved performance and tunable characteristics, making them promising candidates for numerous applications. The experimental setup involves dissolving metal or ferrite target in a solvent (e.g., water, ethanol, chloroform) and irradiating it with a pulsed laser beam. The plasma generates and cools rapidly and condenses to form NPs. The NPs morphology, composition, and size influences by parameters (laser fluence, repetition rate, wavelength, pulse duration) of the target material, and the chemical characteristics of medium. Laser ablation has certain limitations like intense energy concentration that causes thermal damage, affecting substrates. To minimize these effects parameters were optimized. Laser ablation is not suitable for large-scale production, as it has high energy consumption compared to other synthesis protocols that limit its industrial applications. Modern systems are integrated with robotic platforms coupled with fiber-optic allow precise beam steering and control over conditions. This minimizes damage and enhances versatility and microfabrication. Although laser ablation excels at generating ligand-free nanoparticles with uncontaminated surfaces, it faces challenges related to low mass yield and limited scalability for industrial applications [89, 91–99]. Table 2.6 demonstrates the laser ablation synthesis, morphology, and crystal structure of FNMs.

2.2.2 *Ball Milling*

Ball milling is a versatile and adaptable mechanical technique widely employed for particle size reduction, facilitating alloy formation, and synthesizing advanced materials, particularly nanostructures (Fig. 2.6). This technique involves materials and grinding media placed within a rotating or vibrating chamber. The collision between the balls and the powder particles causes mechanical deformation. This continuous interaction refines particle size and triggers solid-state reactions. Ball milling advantages include its simplicity, and the ability to synthesize metastable and nanocrystalline phases at controlled environment [109, 110]. Spinel ferrites can be effectively synthesized via ball milling. Traditional methods for ferrites require high calcination temperatures more than 1000 °C causing undesirable effects like grain growth, inhomogeneity, and low surface area. In contrast, ball milling offers a low-temperature synthesis of ferrites with uniformity and tailored magnetic properties [111, 112].

Ferrites synthesis assisted by ball milling the precursor (F_2O_3, NiO, CoO, ZnO) are mechanically activated by high energy collisions. These collisions introduce lattice defects, enhance surface energy, and diffusion. Prolonged milling durations result in the formation of spinel ferrite. The properties of ferrites synthesized via ball milling are influenced by several parameters like milling time, ball-to-powder weight ratio (BPR), speed, and the type of grinding media. Long milling durations enhance reaction kinetics; however, it introduces undesirable effects like contamination or amorphization. Inert gases like argon or nitrogen prevent oxidation and ensure better control over phase purity. Moreover, milling equipment including planetary, attritor,

Table 2.6 Synthesis of ferrites by the laser ablation method

Sr. no.	Material	Precursors	Laser source	Duration	Pulse width	Wavelength (nm)	Frequency (Hz)	Size (nm)	Fluence (J/cm^2)	Ref
1	$NiFe_2O_4$	NiO, Fe_2O_3	N-IR laser	5 min	45 μm	1064 nm	300 kHz	Particle size = 1–50 nm		[101]
2	$CoFe_2O_4$	Fe_2O_3, Co_3O_4	Nd-YAG laser	10 ns		532 nm	10 Hz		10 J/cm^2	[102]
3	$CoFe_2O_4$	$Co(NO_3)_2.6H_2O$, Fe $(NO_3)_3.9H_2O$	Nd:YAG laser	7 ns		1064 nm	3 Hz	Crystallite size = 15 nm		[103]
4	$Cu_{1-x}Zn_xFe_2O_4$	CuO, ZnO and Fe_2O_3	Nd:YAG laser	10 ns		1064 nm	10 Hz			[104]
5	$BaFe_{12}O_{19}$	$BaCO_3$, cr-Fe_2O_3	KrF excimer laser	23 ns		248 nm	30 Hz	Grain diameter = 400 nm	2 J/cm^2	[105]
6	$SrFe_{12}O_{19}$		KrF excimer laser	20 ns		248 nm	25 Hz	Particle size = 2 μm	2 J/cm^2	[106]
7	Mn–Zn ferrite	$Mn(NO_3)_2{\cdot}6H_2O$, $Zn(NO_3){\cdot}6H_2O$, $Fe(NO_3)_2{\cdot}9H_2O$	KrF excimer laser			248 nm	2.45 GHz	Particle size = 25 nm	8 J/cm^2	[107]
8	$MnFe_2O_4$	MnO, Fe_2O_3	CO_2 laser	50 μs			20 kHz	Particle size = 1–15 nm		[108]

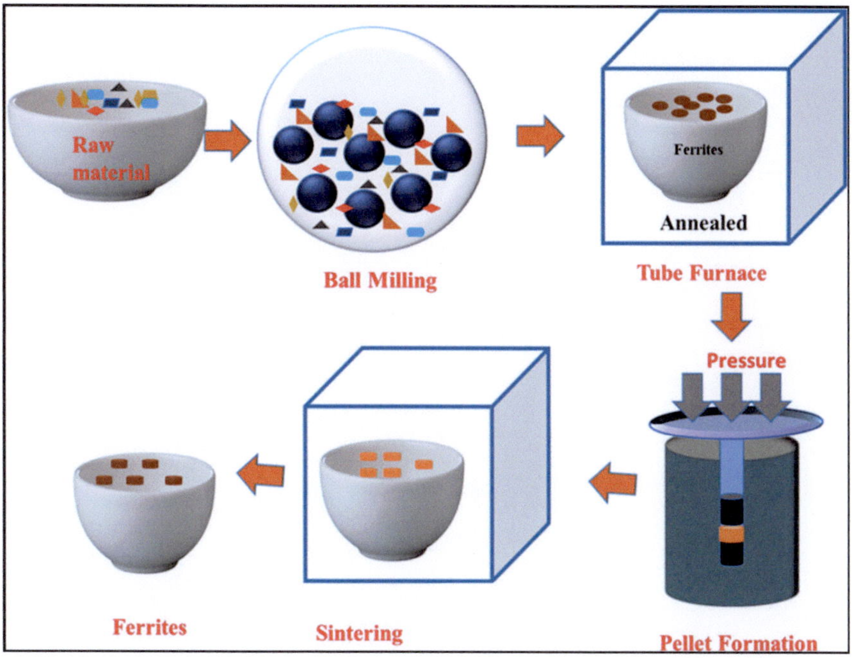

Fig. 2.6 Schematic synthesis of ferrite synthesis via laser ablation. Reproduced with permission from [119]

and SPEX mills, is necessary in determining energy input and consequently effecting the overall efficiency of the reaction [113, 114].

Key advantages include the ability to produce materials with small grain sizes. This refinement improves superparamagnetic behavior, surface area, and magnetic response. These features are important for applications in magnetic storage, catalysis, drug delivery, and electromagnetic interference (EMI) shielding [115–117]. Ball milling offers solvent-free production, ecofriendly synthesis approach. Despite its benefits, it also presents certain limitations including contamination arises from the milling media and achieving complete phase formation without any further thermal treatment. However, after optimizing parameters ball milling proves to be a powerful and efficient technique for fabricating ferrite NPs, making it suitable for a wide range of applications [118]. Table 2.7 represent the ball milling synthesis, morphology, and crystal structure of FNMs.

Table 2.7 Synthesis of ferrites by ball milling method

Sr. no.	Material	Precursors	Milling medium	Milling model	Ball size (mm)	Time	Rpm	Size	Ball to powder ratio	References
1	$ZnFe_2O_4$	ZnO, α-Fe_2O_3	Chrome steel balls	Planetary ball mill	10 mm	30 min–10 h	325 rpm	~6 nm	40:1	[120]
2	$MgFe_2O_4$	MgO, α-Fe_2O_3	Hardened chrome steel ball	Planetary ball mill	10 mm	1–11 h	325 rpm	~4–27 nm	40:1	[121]
3	$BaFe_2O_4$	$BaCO_3$, α-Fe_2O_3				16–62 h	95 rpm	0.1–2 μm	10:1	[114]
4	$CuFe_2O_4$	α-Fe_2O_3, CuO	Hardened steel ball	Planetary ball mill		98 h	950 rpm	Grain size = 6 nm	20:1	[122]
5	Ni-Zn-ferrite	ZnO, NiO, α-Fe_2O_3	Hardened chrome steel balls	High-energy planetary ball mill	10 mm	1–11 h	325 rpm	~4 nm	40:1	[123]
6	$NiFe_2O_4$	$Fe_2Ni_2(OH)_8CO_3 \cdot 2H_2O$, $NiCO_3 \cdot 2Ni(OH)_2 \cdot 4H_2O$	Iron milling balls	Ultrasonic wave-assisted ball milling	1.0–1.5 mm	20 h	235 rpm	Particle size ~ 20 nm	100:1	[124]
7	Cd-Zn ferrite	CdO, ZnO, α-Fe_2O_3	Hardened chrome steel ball	Planetary ball mill	10 mm	25 h	300 rpm	Particle size ~ 4–7 nm	40:1	[125]
8	$MgFe_2O_4$	$4MgCO_3 \cdot Mg(OH)_2 \cdot 5H_2O$ and Fe_2O_3	Stainless steel ball	Ultrasonic wave-assisted ball milling	1.0–1.5 mm	60 h	235 rpm	Particle size ~ 20 nm	100:1	[126]
9	$CuFe_2O_4$	CuO, Fe_2O_3	Stainless steel ball	Planetary ball mill	5 mm	40 h	275 rpm	Diameter = 10 nm	40:1	[127]

2.2.3 *Sputtering*

Sputtering is an extensively employed bottom-up physical vapor deposition (PVD) method for preparing FNMs and thin films (Fig. 2.7) [128, 129]. It involves the ejection of atoms or clusters from a solid target by the energetic bombardment of ions. These atoms or clusters were then condensed onto a substrate creating a thin film. This is a low temperatures process that enables the deposition of highly pure ferrite films with good adhesion and stoichiometric control. The phenomenon was first observed by Grove in 1852, who first observed material from a cathode being deposited as a result of ion bombardment within a gas discharge tube [130]. Modern sputtering uses low-pressure plasma to form ions that collide with material ejecting its atoms. The sputtering setup consists of vacuum chamber, gas supply, source, electrodes, and a substrate holder. This method is effective for ferrite thin films deposition including $NiFe_2O_4$, $CoFe_2O_4$, and $ZnFe_2O_4$ with tailored properties [131–133]. In DC sputtering, a negative voltage is applied to a conductive ferrite target, while the substrate remains grounded. Argon ions strike the target ejecting atoms that deposited onto the substrate. However, DC sputtering is not suitable for conductive or doped targets, accumulate surface charge, disrupting the process. To overcome this limitation, RF sputtering is employed. In this an alternating RF voltage is applied that reverses the target polarity and neutralizes charge buildup, making it suitable for sputtering insulating ferrites [134]. Magnetron sputtering enhances efficiency by applying a magnetic field near the target, trapping electrons and significantly boosting plasma ionization. This enhances deposition rate, uniformity, and firm adhesion that are critical for fabricating ferrite-based devices in spintronics and microwave applications [131, 132]. Magnetron sputtering is compatible with RF and DC depending on target conductivity [135, 136].

Reactive sputtering uses metallic targets (e.g., Fe, Co, Ni) within reactive gas (oxygen or nitrogen) environments, forming ferrite films such as $CoFe_2O_4$ or $NiFe_2O_4$. In contrast, ion beam sputtering uses a focused ion beam to eject material from the target, minimizing interactions and enables controlled film deposition [137]. Several parameters including annealing conditions, target and substrate angle, ion energy and mass, Ar^+ concentration, and substrate temperature influence the properties of the sputtered ferrite film. These parameters determine the film morphology and are crucial for magnetic characteristics including coercivity and saturation magnetization.

Sputtering offers advantages such as strong adhesion, uniform coatings, compositional control, and low-temperature synthesis that makes it suitable for a variety of ferrite synthesis with high structural integrity and magnetic performance. However, limitations involve low deposition rates, small grain sizes, surface roughness, heating, long times, and high equipment costs. Residual gas also affects the film purity [138–140]. Overall, sputtering is an effective method for synthesizing ferrite thin films, essential for various applications. Table 2.8 represents the sputtering methods for the synthesis of FNMs.

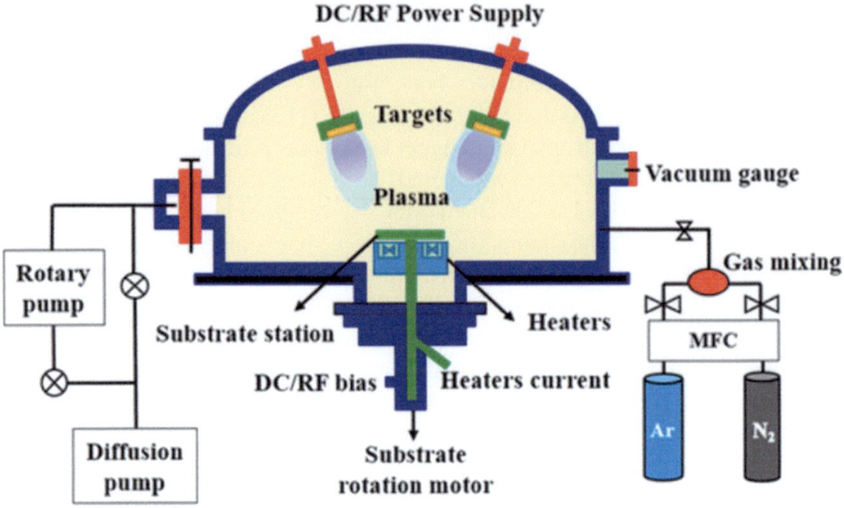

Fig. 2.7 Schematic synthesis illustration by sputtering technique. Reproduced with permission from [141]

2.2.4 *Thermal Evaporation*

Thermal evaporation is a widely employed technique that involves heating a material source under vacuum conditions to form a thin film on a target substrate (Fig. 2.8). This process uses a high DC to energize a resistive element like a resistive coil, boat, or solid bar to reach the required evaporation temperature. This method operates at low-pressure of less than 1024 Pa decreases the contaminants, enables vaporized atoms to travel directly to the substrate without collisions. Due to its simplicity, efficiency, and effectiveness, thermal evaporation continues to be a widely adopted technique in both industrial production and scientific research for thin film fabrication [150–154].

Numerous FNMs have been synthesized through thermal evaporation, including Mg-Zn ferrite [155], $BiFeO_3$ [152, 153], zinc ferrite [156], and cobalt ferrite [157]. In a standard thermal evaporation setup, high-purity (99.99%) stoichiometric powders are placed into an alumina crucible positioned at the center of the heating zone. Substrates left untreated are placed above the source material. The system operates using resistive heating, with temperature precisely regulated by a PID controller and displayed digitally. Deposition time was controlled by mechanical shutter. A mechanical shutter regulates deposition duration. Substrates were cleaned thoroughly by soap solution, acetone, ethanol, and deionized water before deposition to remove surface contaminants [154, 158].

Factors influencing the process are furnace temperature, evaporation duration, source material and the substrate distance. A rotary pump evacuates the chamber

Table 2.8 Synthesis of ferrites via sputtering technique

Sr. no.	Material	Substrate	Gas	Method	Pressure	Size	Ref.
1	$ZnFe_2O_4$ thin film	SiO_2 glass	O_2 and Ar	Radio frequency sputtering techniques	15 mTorr	Crystallite size = 5–20 nm	[131]
2	$NiFe_2O_4$	Borosilicate glass substrates	Ar:O_2	DC Reactive Sputtering Technique	0.8 mbar	Particle size = ~ 10 nm	[132]
3	$ZnFe_2O_4$ thin film	Glass substrate	O_2	RF-magnetron sputtering	27 mTorr	Crystallite size = 8–10 nm	[142]
4	LiZn ferrite	Fused quartz substrates	–	RF-magnetron sputtering	–	Crystallite size = 5 nm	[143]
5	$BaFe_{12}O_{19}$	Alumina and silicon substrates		RF magnetron sputtering	0.4–2.3 P	Grain size = 30 nm	[144]
6	$CoFe_2O_4$ thin film	Si wafers	Ar and O_2	RF magnetron sputtering method	7 mTorr	Grain size = 30–150 nm	[133]
7	Cu–Zn ferrite	Glass substrate	Ar	RF magnetron sputtering method	5–15 mTorr	Particle size = 1.2–3.8 nm	[145]
8	$BiFeO_3$	Si substrate	O_2	Reactive magnetron sputtering	1.33 Pa	Grain size = 80–180 nm	[146]
9	$BaFe_{12}O_{19}$	Si wafers	Ar	RF magnetron sputtering	0.4 and 10 Pa	Grain size = 25–40 nm	[147]
10	MnZn ferrite	Si, MgO, SiO_2/Si	Ar and O_2	RF magnetron sputtering	1.1 Pa	Grain size = 20 nm	[148]
11	$MgFe_2O_4$	Alumina substrate	Ar	RF magnetron sputtering	7 Pa	Crystallite size = 22–48 nm	[149]

to create a low-pressure environment, and oxygen may be added to assist with film formation or oxidation during deposition.

Several advanced variations of thermal evaporation enhance film quality and broaden applicability. These include molecular beam epitaxy (MBE), reactive evaporation, and activated reactive evaporation (ARE). MBE was first introduced in 1960s by Cho and Arthur, enables growth by directing beams onto a heated substrate under high vacuum conditions. It allows control over film thickness and composition. Metals were evaporates by reactive evaporation with reactive gas, resulting in stoichiometric oxides or metal alloys [159, 160]. ARE use plasma to ionize the reactive gas and evaporated atoms increasing reactivity, adhesion, and deposition.

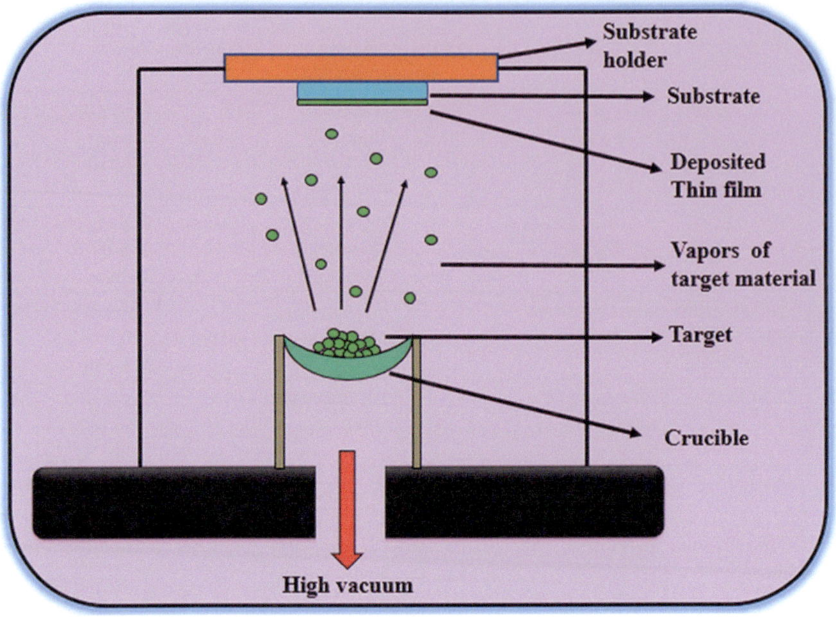

Fig. 2.8 Schematic synthesis diagram by thermal evaporation method. Reproduced with permission from [165]

A negative bias voltage on the substrate further enhances ion interaction with the surface [161, 162]. It is effective for FNMs synthesis with enhanced magnetic and electrical properties that supports the formation of complex nanostructures including nanotubes and hollow particles [152].

This technique controls film thickness, uniform deposition of ultrathin layers, and formation of multilayer structures. It enables localized deposition with shutters and masks making it suitable for low melting point materials. This methods has limitations like contamination and difficulties in maintaining stoichiometric balance [163, 164]. Alloy combinations can be complex and selecting chemically inert and thermally stable boat materials adds to the challenges.

References

1. Pund, S.N., et al.: Preparation techniques for zinc ferrites and their applications: a review. Mater. Today: Proc. **60**, 2194–2208 (2022)
2. Houshiar, M., et al.: Synthesis of cobalt ferrite (CoFe2O4) nanoparticles using combustion, coprecipitation, and precipitation methods: a comparison study of size, structural, and magnetic properties. J. Magn. Magn. Mater. **371**, 43–48 (2014)
3. Massart, R.: Preparation of aqueous magnetic liquids in alkaline and acidic media. IEEE Trans. Magn. **17**(2), 1247–1248 (1981)

4. Mirzaee, S., Azizian-Kalandaragh, Y., Rahimzadeh, P.: Modified co-precipitation process effects on the structural and magnetic properties of Mn-doped nickel ferrite nanoparticles. Solid State Sci. **99**, 106052 (2020)
5. Ram, J., et al.: Effect of annealing on the surface morphology, optical and structural properties of nanodimensional tungsten oxide prepared by coprecipitation technique. J. Electron. Mater. **48**, 1174–1183 (2019)
6. Anik, M.I., et al.: Recent progress of magnetic nanoparticles in biomedical applications: a review. Nano Select **2**(6), 1146–1186 (2021)
7. Soufi, A., et al.: Spinel ferrites nanoparticles: Synthesis methods and application in heterogeneous Fenton oxidation of organic pollutants–a review. Appl. Surf. Sci. Adv. **6**, 100145 (2021)
8. Ramadoss, G., et al.: Mesoporous ferromagnetic manganese ferrite nanoparticles for enhanced visible light mineralization of azoic dye into nontoxic by-products. Sci. Total. Environ. **765**, 142707 (2021)
9. Kim, Y.I., Kim, D., Lee, C.S.: Synthesis and characterization of CoFe2O4 magnetic nanoparticles prepared by temperature-controlled coprecipitation method. Phys. B **337**(1–4), 42–51 (2003)
10. Shahraki, R.R., et al.: Structural characterization and magnetic properties of superparamagnetic zinc ferrite nanoparticles synthesized by the coprecipitation method. J. Magn. Magn. Mater. **324**(22), 3762–3765 (2012)
11. Joshi, S., et al.: Structural, magnetic, dielectric and optical properties of nickel ferrite nanoparticles synthesized by co-precipitation method. J. Mol. Struct. **1076**, 55–62 (2014)
12. Maaz, K., et al.: Synthesis and magnetic characterization of nickel ferrite nanoparticles prepared by co-precipitation route. J. Magn. Magn. Mater. **321**(12), 1838–1842 (2009)
13. Suharyadi, E., Hermawan, A., Puspitarum, D.: Crystal structure and magnetic properties of magnesium ferrite (MgFe2O4) nanoparticles synthesized by coprecipitation method. J. Phys. Conf. Ser. IOP Publishing (2018)
14. Agouriane, E., et al.: Structural and magnetic properties of CuFe2O4 ferrite nanoparticles synthesized by co-precipitation. J. Mater. Environ. Sci **7**(11), 4116–4120 (2016)
15. Sivanandan, T., Saravanan, S.: Effects of calcination temperatures on structural, functional, morphological, and magnetic properties of strontium ferrite (SrFe2O4) nanoparticles. Kuwait J. Sci. **50**(3), 271–275 (2023)
16. Sedaghati-Jamalabad, G., Bagheri-Mohagheghi, M.: Influence of synthesis route on structural properties of SnFe2O4 spinel phase via methods of co-precipitation, sol–gel and solvothermal: morphology, phase analysis, crystallite size and lattice strain. Discov. Appl. Sci. **6**(4), 202 (2024)
17. Maleki, H., Haselpour, M., Fathi, R.: The effect of calcination conditions on structural and magnetic behavior of bismuth ferrite synthesized by co-precipitation method. J. Mater. Sci. Mater. Electron. **29**, 4320–4326 (2018)
18. Lestari, M., Taufiq, A., Hidayat, A.: Green synthesis of CrFe2O4 nanoparticles using Cucumis sativus as a natural surfactant. Mater. Today Proc. **44**, 3221–3224 (2021)
19. Bokov, D., et al.: Nanomaterial by sol-gel method: synthesis and application. Adv. Mater. Sci. Eng. **2021**(1), 5102014 (2021)
20. Hench, L.L., West, J.K.: The sol-gel process. Chem. Rev. **90**(1), 33–72 (1990)
21. Vinosha, P.A., et al.: Review on recent advances of zinc substituted cobalt ferrite nanoparticles: synthesis characterization and diverse applications. Ceram. Int. **47**(8), 10512–10535 (2021)
22. Sharma, K., et al.: A review paper: synthesis techniques and advance application of Mn-Zn nano-ferrites. Mater. Today Proc. (2023)
23. Jabbar, R., Sabeeh, S.H., Hameed, A.M.: Structural, dielectric and magnetic properties of Mn+ 2 doped cobalt ferrite nanoparticles. J. Magn. Magn. Mater. **494**, 165726 (2020)
24. Zayat, M., Levy, D.: Blue CoAl2O4 particles prepared by the sol−gel and citrate−gel methods. Chem. Mater. **12**(9), 2763–2769 (2000)
25. Hilpert, R., Schweinhagen, R.: Über Ferrite. (III. Mitteilung). Z. Phys. Chem. **31**(1) 1–11 (1936)

26. Tavakoli, A., Sohrabi, M., Kargari, A.: A review of methods for synthesis of nanostructured metals with emphasis on iron compounds. Chem. Pap. **61**(3), 151–170 (2007)
27. Biswas, B., et al.: Investigation of structural, optical, and magnetic properties of NiFe2O4 for efficient photocatalytic degradation of organic pollutants through photo fenton reactions. Heliyon **10**(17) (2024)
28. Hussain, A., Abbas, T., Niazi, S.B.: Preparation of Ni1− xMnxFe2O4 ferrites by sol–gel method and study of their cation distribution. Ceram. Int. **39**(2), 1221–1225 (2013)
29. Thorat, L., et al.: Ni 2+-substituted Mg–Cu–Zn ferrites: a colloidal approach of tuning structural and electromagnetic properties. J. Sol-Gel Sci. Technol. **86**, 731–742 (2018)
30. Sajjia, M., et al.: Developments of cobalt ferrite nanoparticles prepared by the sol–gel process. Ceram. Int. **40**(1), 1147–1154 (2014)
31. Chihi, I., et al.: Sol–gel synthesis and characterization of magnesium ferrites by XRD, TEM, EPR, Mossbauer, and impedance spectroscopy. J. Mater. Sci. Mater. Electron. **32**, 16634–16647 (2021)
32. Zhang, L., Wu, Y.: Sol-gel synthesized magnetic MnFe2O4 spinel ferrite nanoparticles as novel catalyst for oxidative degradation of methyl orange. J. Nanomater. **2013**(1), 640940 (2013)
33. Peymanfar, R., et al.: Preparation and identification of BaFe2O4 nanoparticles by the sol–gel route and investigation of its microwave absorption characteristics at ku-band frequency using silicone rubber medium. In: Proceedings. MDPI (2018)
34. Renuka, L., et al.: Synthesis of ZnFe2O4 nanoparticle by combustion and sol gel methods and their structural, photoluminescence and photocatalytic performance. Mater. Today Proc. **5**(10), 20819–20826 (2018)
35. Sen, R., et al.: Synthesis and characterization of nickel ferrite (NiFe2O4) nanoparticles prepared by sol-gel method. Mater. Today Proc. **2**(4–5), 3750–3757 (2015)
36. Ghahfarokhi, S.M., Shobegar, E.M., Shoushtari, M.Z.: Preparation and characterization of spinel SrFe2O4 nanoparticles by method sol-gel. J. Aust. Ceram. Soc. **57**(5), 1359–1369 (2021)
37. Ghahfarokhi, S.M., Shobegar, E.M.: Influence of pH on the structural, magnetic and optical properties of SrFe2O4 nanoparticles. J. Mater. Res. Technol. **9**(6), 12177–12186 (2020)
38. Soufi, A., et al.: Heterogeneous Fenton-like degradation of tartrazine using CuFe2O4 nanoparticles synthesized by sol-gel combustion. Appl. Surf. Sci. Adv. **9**, 100251 (2022)
39. Wu, W., et al.: Recent progress on magnetic iron oxide nanoparticles: synthesis, surface functional strategies and biomedical applications. Sci. Technol. Adv. Mater. **16**(2), 023501 (2015)
40. Sarkar, K., et al.: Cation vacancy and magnetic properties of ZnFe2O4 microspheres. Phys. B **583**, 412015 (2020)
41. Shaterian, M., Rezvani, A., Abbasian, A.R.: Controlled synthesis and self-assembly of ZnFe2O4 nanoparticles into microspheres by solvothermal method. Mater. Res. Express. **6**(12), 1250e5 (2020)
42. Kalam, A., et al.: Modified solvothermal synthesis of cobalt ferrite (CoFe2O4) magnetic nanoparticles photocatalysts for degradation of methylene blue with H2O2/visible light. Results Phys. **8**, 1046–1053 (2018)
43. Wang, J., et al.: Solvothermal synthesis and magnetic properties of size-controlled nickel ferrite nanoparticles. J. Alloy. Compd. **479**(1–2), 791–796 (2009)
44. Huo, Y., et al.: Solvothermal synthesis and applications of micro/nano carbons: a review. Chem. Eng. J. **451**, 138572 (2023)
45. Rani, K., et al.: Improved thermoelectric performance of Se-doped n-type nanostructured Bi2Te3. J. Mater. Sci. Mater. Electron. **34**(13), 1074 (2023)
46. Duong, H.D.T., Nguyen, D.T., Kim, K.-S.: Effects of process variables on properties of CoFe2O4 nanoparticles prepared by solvothermal process. Nanomaterials **11**(11), 3056 (2021)
47. Shen, Y., et al.: One-pot synthesis of MgFe2O4 nanospheres by solvothermal method. Mater. Lett. **96**, 85–88 (2013)

48. Zhang, Z., et al.: Facile solvothermal synthesis of mesoporous manganese ferrite (MnFe2O4) microspheres as anode materials for lithium-ion batteries. J. Colloid Interface Sci. **398**, 185–192 (2013)
49. Liu, R., et al.: Solvothermal synthesis of size-tunable ZnFe2O4 colloidal nanocrystal assemblies and their electrocatalytic activity towards hydrogen peroxide. J. Magn. Magn. Mater. **424**, 155–160 (2017)
50. Wang, J., et al.: Solvothermal synthesis and characterization of NiFe2O4 nanospheres with adjustable sizes. Solid State Commun. **150**(25–26), 1141–1144 (2010)
51. Seetharaman, D.: Investigation of thermal stability, structure, magnetic and dielectric properties of solvothermally synthesised SnFe2O4. Open Ceram. **9**, 100222 (2022)
52. Yang, X., et al.: Solvothermal synthesis of porous CuFe2O4 nanospheres for high performance acetone sensor. Sens. Actuators B Chem. **270**, 538–544 (2018)
53. Sabnis, S., Bhadane, P.A., Kulkarni, P.: Process flow of spray pyrolysis technique. J. Appl. Phys. **4**(5), 7–11 (2013)
54. Mooney, J.B., Radding, S.B.: Spray pyrolysis processing. Annu. Rev. Mater. Sci. **12**(1), 81–101 (1982)
55. Lee, C., Chang, H., Jang, H.D.: Preparation of CoFe2O4-graphene composites using aerosol spray pyrolysis for supercapacitors application. Aerosol Air Qual. Res. **19**(3), 443–448 (2019)
56. Grace, J., Marijnissen, J.: A review of liquid atomization by electrical means. J. Aerosol Sci. **25**(6), 1005–1019 (1994)
57. Tina, S., Sudha Kartha, C., Vijayakumar, K.: Automation of chemical spray pyrolysis unit and fabrication of sprayed CuInS2/In2S3 solar cell. Cochin University of Science & Technology (2009)
58. Lorentzou, S., Agrafiotis, C.C., Konstandopoulos, A.G.: Aerosol spray pyrolysis synthesis of water-splitting ferrites for solar hydrogen production. Granul. Matter **10**(2), 113–122 (2008)
59. Eslamian, M., Ahmed, M., Ashgriz, N.: Modelling of nanoparticle formation during spray pyrolysis. Nanotechnology **17**(6), 1674 (2006)
60. Mädler, L., et al.: Controlled synthesis of nanostructured particles by flame spray pyrolysis. J. Aerosol Sci. **33**(2), 369–389 (2002)
61. Jundale, V.A., Chorage, G.Y., Yadav, A.A.: Structural, morphological and optical properties of CoFe2O4 thin film by spray pyrolysis technique. Materials Today: Proceedings **43**, 2678–2681 (2021)
62. Das, H., et al.: Impact of precursor solution concentration to form superparamagnetic MgFe2O4 nanospheres by ultrasonic spray pyrolysis technique for magnetic thermotherapy. Adv. Powder Technol. **28**(7), 1696–1703 (2017)
63. Ozdemir, E.T., et al.: Effects of annealing on the structural and magnetic properties of flame spray pyrolyzed MnFe2O4 nanoparticles. Int. J. Appl. Ceram. Technol. **20**(6), 3775–3785 (2023)
64. Chikundre, C., Birajdar, A.: Spray pyrolysis deposition and characterization of nanocrystalline MnFe2O4 thin film. Int. J. Adv. Innov. Res. **6**(1), 25–28 (2019)
65. Yildirim, S., et al.: The structural, magnetic, and optical properties of flame spray pyrolysis-derived spinel NiFe2O4 nanoparticles. J. Am. Ceram. Soc. **108**(2), e20220 (2025)
66. Lv, L., et al.: Ultrasonic spray pyrolysis synthesis of three-dimensional ZnFe2O4-based macroporous spheres for excellent sensitive acetone gas sensor. Sens. Actuators B Chem. **297**, 126755 (2019)
67. Lin, Z., et al.: Elucidation of the optical, electronic, and photoelectrochemical properties of p-type CuFe2O4 photocathodes. Int. J. Hydrog. Energy **77**, 511–519 (2024)
68. Lee, H.M., et al.: Preparation of unsintered spherical BaFe12O19 nanoparticles via an alcohol-assisted spray-pyrolysis route. J. Am. Ceram. Soc. **91**(9), 2856–2861 (2008)
69. Jamkhande, P.G., et al.: Metal nanoparticles synthesis: an overview on methods of preparation, advantages and disadvantages, and applications. J. Drug Deliv. Sci. Technol. **53**, 101174 (2019)
70. Yang, G., Park, S.-J.: Conventional and microwave hydrothermal synthesis and application of functional materials: a review. Materials **12**(7), 1177 (2019)

71. Sasikala, S.P., Poulin, P., Aymonier, C.: Advances in subcritical hydro-/solvothermal processing of graphene materials. Adv. Mater. **29**(22), 1605473 (2017)
72. Casbeer, E., Sharma, V.K., Li, X.-Z.: Synthesis and photocatalytic activity of ferrites under visible light: a review. Sep. Purif. Technol. **87**, 1–14 (2012)
73. Yao, L., et al.: Synthesis of cobalt ferrite with enhanced magnetostriction properties by the sol– gel– hydrothermal route using spent Li-ion battery. J. Alloy. Compd. **680**, 73–79 (2016)
74. Chen, D., Xu, R.: Hydrothermal synthesis and characterization of nanocrystalline Fe3O4 powders. Mater. Res. Bull. **33**(7), 1015–1021 (1998)
75. Burda, C., et al.: Chemistry and properties of nanocrystals of different shapes. Chem. Rev. **105**(4), 1025–1102 (2005)
76. Sokola, P., et al.: Comprehensive study of stereolithography and digital light processing printing of zirconia photosensitive suspensions. Ceramics **7**(4), 1616–1638 (2024)
77. Zhu, J., et al.: Progress in the preparation and modification of zinc ferrites used for the photocatalytic degradation of organic pollutants. Int. J. Environ. Res. Public Health **19**(17), 10710 (2022)
78. Ji, G., et al.: Simplified synthesis of single-crystalline magnetic CoFe2O4 nanorods by a surfactant-assisted hydrothermal process. J. Cryst. Growth **270**(1–2), 156–161 (2004)
79. Reddy, M.P., et al.: A facile hydrothermal synthesis, characterization and magnetic properties of mesoporous CoFe2O4 nanospheres. J. Magn. Magn. Mater. **388**, 40–44 (2015)
80. Tahir, M., et al.: Phase formation and dielectric properties of MgFe2O4 nanoparticles synthesized by hydrothermal technique. Heliyon **10**(8) (2024)
81. Idris, N., Mustafa, N., Ismail, M.: MnFe2O4 nanopowder synthesised via a simple hydrothermal method for promoting hydrogen sorption from MgH2. Int. J. Hydrog. Energy **42**(33), 21114–21120 (2017)
82. Han, L., et al.: Synthesis of ZnFe2O4 nanoplates by succinic acid-assisted hydrothermal route and their photocatalytic degradation of rhodamine B under visible light. J. Environ. Chem. Eng. **2**(1), 123–130 (2014)
83. Samson, V.A., et al.: One pot hydrothermal synthesis and characterization of NiFe2O4 nanoparticles. Mater. Today Proc. **50**, 2665–2667 (2022)
84. Sargazi, S., et al.: Assessment of snfe2o4 nanoparticles for potential application in theranostics: synthesis, characterization, in vitro, and in vivo toxicity. Materials **14**(4), 825 (2021)
85. Mandizadeh, S., Salavati-Niasari, M., Sadri, M.: Hydrothermal synthesis, characterization and magnetic properties of BaFe2O4 nanostructure as a photocatalytic oxidative desulfurization of dibenzothiophene. Sep. Purif. Technol. **175**, 399–405 (2017)
86. Zapukhlyak, R., et al.: Structure and magnetic properties of hydrothermally synthesized CuFe2O4 and CuFe2O4/rGO composites. J. Magn. Magn. Mater. **587**, 171208 (2023)
87. Chandranna, S.N., et al.: Hydrothermal synthesis of BiFe2O4 heterostructure for photodegradation of dye and biological implications. Environ. Surf. Interfaces **3**, 24–32 (2025)
88. Kim, M., et al.: Synthesis of nanoparticles by laser ablation: a review. Kona Powder Part. J. **34**, 80–90 (2017)
89. Brown, M.S., Arnold, C.B.: Fundamentals of laser-material interaction and application to multiscale surface modification. In: Laser Precision Microfabrication, pp. 91–120. Springer, Berlin (2010)
90. Ravi-Kumar, S., et al.: Laser ablation of polymers: a review. Polym. Int. **68**(8), 1391–1401 (2019)
91. Sportelli, M.C., et al.: The pros and cons of the use of laser ablation synthesis for the production of silver nano-antimicrobials. Antibiotics **7**(3), 67 (2018)
92. Amendola, V., Meneghetti, M.: Laser ablation synthesis in solution and size manipulation of noble metal nanoparticles. Phys. Chem. Chem. Phys. **11**(20), 3805–3821 (2009)
93. Shukla, A., et al.: Laser synthesized magnetically recyclable titanium ferrite nanoparticles for photodegradation of dyes. J. Mater. Sci. Mater. Electron. **28**(20), 15380–15386 (2017)
94. Duque, J.S., et al.: Colloidal metal oxide nanoparticles prepared by laser ablation technique and their antibacterial test. Colloids Interfaces **3**(1), 25 (2019)

95. Kim, B., et al.: Selective laser ablation of metal thin films using ultrashort pulses. Int. J. Precis. Eng. Manuf.-Green Technol. **8**(3), 771–782 (2021)
96. Leong, K., et al.: Laser-based characterization and decontamination of contaminated facilities. In: International Congress on Applications of Lasers & Electro-Optics. Laser Institute of America (1996)
97. Almessiere, M.A., et al.: Structural and magnetic properties of Co0. 5Ni0. 5Ga0. 01Gd0. 01Fe1. 98O4/ZnFe2O4 spinel ferrite nanocomposites: comparative study between sol-gel and pulsed laser ablation in liquid approaches. Nanomaterials **11**(9), 2461 (2021)
98. Fazio, E., et al.: Nanoparticles engineering by pulsed laser ablation in liquids: concepts and applications. Nanomaterials **10**(11), 2317 (2020)
99. Almessiere, M., et al.: Morphological, structural, and magnetic characterizations of hard-soft ferrite nanocomposites synthesized via pulsed laser ablation in liquid. Mater. Sci. Eng. B **273**, 115446 (2021)
100. Shukla, A., et al.: Target phase-induced compositional control in liquid-phase pulsed laser ablation produced titanium ferrite nanomaterials. Bull. Mater. Sci. **44**(2), 152 (2021)
101. Özçelik, S., et al.: Structure, magnetic, photocatalytic and blood compatibility studies of nickel nanoferrites prepared by laser ablation technique in distilled water. J. Alloy. Compd. **854**, 157279 (2021)
102. Bulai, G., et al.: Pure and rare earth doped cobalt ferrite laser ablation: space and time resolved optical emission spectroscopy. Dig. J. Nanomater. Biostruct. **10**(3), 1043–1053 (2015)
103. Ismail, M.M., et al.: Influence of laser ablation energy on structural and optical properties of CoLa0. 02Fe1. 98O4 nanoferrite for bioactivity application. J. Mater. Eng. Perform. **34**(10), 8841–8847 (2025)
104. Ahmed, M., Khalil, A., Solyman, S.: Laser induced structural and transport properties change in Cu–Zn ferrites. J. Mater. Sci. **42**(11), 4098–4109 (2007)
105. Masterson, H., et al.: Thin films of barium ferrite with perpendicular magnetic anisotropy produced by laser ablation deposition. J. Appl. Phys. **73**(8), 3917–3921 (1993)
106. Nawathey-Dikshit, R., et al.: Synthesis of single domain strontium ferrite powder by pulsed laser ablation. Appl. Phys. Lett. **68**(24), 3491–3493 (1996)
107. Praveena, K., Sadhana, K., Murthy, S.: Elastic behaviour and internal friction studies on nanocrystalline Mn–Zn ferrite films prepared by the method of pulsed laser ablation. J. Alloy. Compd. **492**(1–2), 245–250 (2010)
108. Özçelik, S., et al.: Properties and antibacterial activity of MnFe2O4 nanoparticles obtained by pulsed laser ablation in liquid. Open Ceram. **17**, 100555 (2024)
109. Suryanarayana, C.: Mechanical alloying and milling. Prog. Mater. Sci. **46**(1–2), 1–184 (2001)
110. Abdellaoui, M., Gaffet, E.: The physics of mechanical alloying in a planetary ball mill: mathematical treatment. Acta Metall. Mater. **43**(3), 1087–1098 (1995)
111. Bid, S., Sahu, P., Pradhan, S.: Microstructure characterization of mechanosynthesized nanocrystalline NiFe2O4 by Rietveld's analysis. Physica E **39**(2), 175–184 (2007)
112. Todaka, Y., et al.: Synthesis of ferrite nanoparticles by mechanochemical processing using a ball mill. Mater. Trans. **44**(2), 277–284 (2003)
113. Mahdikhah, V., et al.: Control of structural and magnetic characteristics of cobalt ferrite by post-calcination mechanical milling. J. Phys. Chem. Solids **134**, 286–294 (2019)
114. Mendoza-Suarez, G., et al.: Magnetic properties and microstructure of Ba-ferrite powders prepared by ball milling. J. Magn. Magn. Mater. **223**(1), 55–62 (2001)
115. Bergheul, S., et al.: Magnetic microwave and absorbing properties of Fe-Co alloy synthesised by mechanical alloying process. Int. J. Microstruct. Mater. Prop. **1**(3–4), 334–340 (2006)
116. Syazwan, M., et al.: Co–Ti-and Mn–Ti-substituted barium ferrite for electromagnetic property tuning and enhanced microwave absorption synthesized via mechanical alloying. J. Aust. Ceram. Soc. **53**(2), 465–474 (2017)
117. Moravvej-Farshi, F., Amishi, M., Nekouee, K.A.: Influence of different milling time on synthesized Ni–Zn ferrite properties by mechanical alloying method. J. Mater. Sci. Mater. Electron. **31**(16), 13610–13619 (2020)

118. Gui-yin, Z., et al.: Application of mechanical ball-milling technology in material preparation. Powder Metall. Technol. **36**(4), 315–318 (2018)
119. Thakur, P., et al.: Mg-Zn Ferrite nanoparticles: a brief review on synthesis, characterizations, and applications. ChemBioEng Rev. **10**(5), 711–736 (2023)
120. Bid, S., Pradhan, S.: Preparation of zinc ferrite by high-energy ball-milling and microstructure characterization by Rietveld's analysis. Mater. Chem. Phys. **82**(1), 27–37 (2003)
121. Pradhan, S., et al.: Microstructure characterization and cation distribution of nanocrystalline magnesium ferrite prepared by ball milling. Mater. Chem. Phys. **93**(1), 224–230 (2005)
122. Goya, G., Rechenberg, H., Jiang, J.: Structural and magnetic properties of ball milled copper ferrite. J. Appl. Phys. **84**(2), 1101–1108 (1998)
123. Bid, S., Pradhan, S.: Characterization of crystalline structure of ball-milled nano-Ni–Zn-ferrite by Rietveld method. Mater. Chem. Phys. **84**(2–3), 291–301 (2004)
124. Chen, D., Liu, H.-Y.: One-step synthesis of nickel ferrite nanoparticles by ultrasonic wave-assisted ball milling technology. Mater. Lett. **72**, 95–97 (2012)
125. Sinha, M., Pradhan, S.: Synthesis of nanocrystalline Cd–Zn ferrite by ball milling and its stability at elevated temperatures. J. Alloy. Compd. **489**(1), 91–98 (2010)
126. Chen, D., Li, D.-Y., Kang, Z.-T.: Preparation of magnesium ferrite nanoparticles by ultrasonic wave-assisted aqueous solution ball milling. Ultrason. Sonochem. **20**(6), 1337–1340 (2013)
127. Cobos, M.Á., et al.: Ball milling and annealing effect in structural and magnetic properties of copper ferrite by ceramic synthesis. J. Alloy. Compd. **1006**, 176206 (2024)
128. Wender, H., et al.: Sputtering deposition of nanoparticles onto liquid substrates: recent advances and future trends. Coord. Chem. Rev. **257**(17–18), 2468–2483 (2013)
129. El-Khawaga, A.M., Zidan, A., Abd El-Mageed, A.I.: Preparation methods of different nanomaterials for various potential applications: a review. J. Mol. Struct. **1281**, 135148 (2023)
130. Grove, W.R., VII: On the electro-chemical polarity of gases. Philos. Trans. R. Soc. Lond. **142**, 87–101 (1852)
131. Nakashima, S., et al.: Effect of heat treatment on the magnetic properties of zinc ferrite thin film prepared by a sputtering method. J. Ceram. Soc. Jpn. **12-1**(PacRim5 Special Issue) (2004)
132. Naji, N.E., Aljubouri, A.A., Ismail, R.A.: Synthesis and characterization of nickel ferrite nanostructures by DC reactive sputtering technique using new target configuration. Plasmonics **20**(4), 2003–2011 (2025)
133. Lee, J.-G., Chae, K.P., Sur, J.C.: Surface morphology and magnetic properties of CoFe2O4 thin films grown by a RF magnetron sputtering method. J. Magn. Magn. Mater. **267**(2), 161–167 (2003)
134. Khan, F.A.: Synthesis of nanomaterials: methods & technology. In: Applications of Nanomaterials in Human Health, pp. 15–21. Springer, Berlin (2020)
135. Zagho, M.M., et al.: A brief overview of RF sputtering deposition of boron carbon nitride (BCN) thin films. Emerg. Mater. **2**(1), 79–93 (2019)
136. Wang, Y.-H., et al.: A review on the pathways of the improved structural characteristics and photocatalytic performance of titanium dioxide (TiO2) thin films fabricated by the magnetron-sputtering technique. Catalysts **10**(6), 598 (2020)
137. Becker, M., et al.: Materials processing using radio-frequency ion-sources: Ion-beam sputter-deposition and surface treatment. Rev. Sci. Instrum. **90**(2), 023901 (2019)
138. Ma, Y., et al.: Materials and structure engineering by magnetron sputtering for advanced lithium batteries. Energy Stor. Mater. **39**, 203–224 (2021)
139. Lee, J.-H., Lee, D.-J.: Effects of CdCl2 treatment on the properties of CdS films prepared by rf magnetron sputtering. Thin Solid Films **515**(15), 6055–6059 (2007)
140. Hadis, N.S.M., Abd Manaf, A., Herman, S.H.: Comparison on TiO2 thin film deposition method for fluidic based glucose memristor sensor. In: 2015 IEEE International Circuits and Systems Symposium (ICSyS). IEEE (2015)
141. Khan, W., et al.: The effect of sputtering parameters and doping of copper on surface free energy and magnetic properties of iron and iron nitride nano thin films on polymer substrate. Materials **10**(2), 217 (2017)

142. Sultan, M., Singh, R.: Magnetic and optical properties of RF-sputtered zinc ferrite thin films. J. Appl. Phys. **105**(7), 07A512 (2009)
143. Dash, J., et al.: Study of magnetization and crystallization in sputter deposited LiZn ferrite thin films. J. Appl. Phys. **86**(6), 3303–3311 (1999)
144. Capraro, S., et al.: Properties of barium ferrite sputtered films. Sens. Actuators A **113**(3), 382–386 (2004)
145. Sultan, M., Singh, R.: Crystal structure and magnetic properties of RF-sputtered Cu–Zn ferrite thin films. J. Appl. Phys. **107**(9), 09A510 (2010)
146. Iljinas, A., Stankus, V.: Structural and ferroelectric properties of bismuth ferrite thin films deposited by direct current reactive magnetron sputtering. Thin Solid Films **601**, 106–110 (2016)
147. Capraro, S., et al.: Crystallographic properties of magnetron sputtered barium ferrite films. Mater. Sci. Eng. B **112**(1), 19–24 (2004)
148. Wang, L., et al.: The influence of substrate on the magnetic properties of MnZn ferrite thin film fabricated by alternate sputtering. Phys. Status Solidi (A) **205**(10), 2453–2457 (2008)
149. Kotnala, R., et al.: Influence of annealing on humidity response of RF sputtered nanocrystalline MgFe2O4 thin films. Thin Solid Films **519**(18), 6135–6139 (2011)
150. Abegunde, O.O., et al.: Overview of thin film deposition techniques. AIMS Mater. Sci. **6**(2), 174–199 (2019)
151. Shi, Q., et al.: Structural, optical and photoluminescence properties of Ga2O3 thin films deposited by vacuum thermal evaporation. J. Lumin. **206**, 53–58 (2019)
152. Manzoor, A., et al.: Synthesis and characterization of Bismuth ferrite (BiFeO3) nanoparticles by solution evaporation method. J. Magn. Magn. Mater. **393**, 269–272 (2015)
153. Afzal, A., et al.: Effect of O-vacancies on magnetic properties of bismuth ferrite nanoparticles by solution evaporation method. J. Magn. Magn. Mater. **399**, 77–80 (2016)
154. Basak, A., et al.: Effect of substrate on the structural, optical and electrical properties of SnS thin films grown by thermal evaporation method. Thin Solid Films **645**, 97–101 (2018)
155. Mezher, S.J., et al.: Microwave properties of Mg–Zn ferrite deposited by the thermal evaporation technique. Vacuum **173**, 109114 (2020)
156. Naseri, M.G., et al.: Synthesis and characterization of zinc ferrite nanoparticles by a thermal treatment method. Solid State Commun. **151**(14–15), 1031–1035 (2011)
157. Goodarz Naseri, M., et al.: Simple synthesis and characterization of cobalt ferrite nanoparticles by a thermal treatment method. J. Nanomater. **2010**(1), 907686 (2010)
158. Keshav, R., Mahesha, M.: Investigation on performance of CdTe solar cells with CdS and bilayer ZnS/CdS windows grown by thermal evaporation technique. Int. J. Energy Res. **45**(5), 7421–7435 (2021)
159. Pulker, H.K., Paesold, G., Ritter, E.: Refractive indices of TiO2 films produced by reactive evaporation of various titanium–oxygen phases. Appl. Opt. **15**(12), 2986–2991 (1976)
160. Mergel, D., et al.: Density and refractive index of TiO2 films prepared by reactive evaporation. Thin Solid Films **371**(1–2), 218–224 (2000)
161. Bunshah, R., Raghuram, A.: Activated reactive evaporation process for high rate deposition of compounds. J. Vac. Sci. Technol. **9**(6), 1385–1388 (1972)
162. Bunshah, R.: The activated reactive evaporation process: developments and applications. Thin Solid Films **80**(1–3), 255–261 (1981)
163. Mattox, D.M.: Physical vapor deposition (PVD) processes. Met. Finish. **100**, 394–408 (2002)
164. Wang, S., et al.: Fabrication of efficient metal halide perovskite solar cells by vacuum thermal evaporation: a progress review. Curr. Opin. Electrochem. **11**, 130–140 (2018)
165. Abid, N., et al.: Synthesis of nanomaterials using various top-down and bottom-up approaches, influencing factors, advantages, and disadvantages: a review. Adv. Coll. Interface. Sci. **300**, 102597 (2022)

Chapter 3
Characterization of Ferrites

3.1 Structural Properties

The crystal structure and lattice constant of materials significantly influence their characteristics. In crystalline materials, factors such as crystallite size, lattice strain, surface area, crystallinity and porosity delineate the scope of their applicability. XRD is a technique employed to ascertain the structure of various materials. Dora et al. [1] determined 8.369 and 8.260 Å lattice parameters of pristine and modified Mg-Zn ferrite with crystallite size reduction from 14.7 to 13 nm. This reduction was anticipated because of the substitution of larger-radius zinc ions with smaller-radius Mg^{2+} ions. Ravikumar et al. [2] indicated that the crystallite size increased from 25 to 35 nm, and the lattice constant, which varied from 8.407 to 8.356 Å, exhibited a linear decline with the increasing ion concentration in Mg-Zn, demonstrating adherence to Vegard's rule. Singh et al. [3] observed that the lattice parameter rose with larger Mg^{2+} concentration, reaching a maximum at $x = 0.6$, after which it declined. This resulted from the influence of the secondary phase in the Mg-rich environment. The crystallite diameters ranged from 6 to 14 nm, exhibiting a nonlinear fluctuation. Doping substantially influences the fundamental structure of crystals due to variations in ionic radii, the development of bond energy, and other factors. Thakur et al. [4] identified the cubic phase with crystallite size increased from 60.97 to 70.13 nm after doping (Fig. 3.1a and b). Somvanshi et al. reported cubic phase of Gd modified Zn-Mg ferrites with enhancement in crystallite size from 21.04 to 23.35 nm. With a higher amount of Gd^{3+}, crystallite size was reduced and X-ray density increased [4].

Ramiza et al. [5] obtained a smaller size (6.198) of Mn-Zn ferrites under UV light illumination. The smaller crystallite size was beneficial for many applications owing to its larger surface area. Thakur et al. [6] investigated the enhancement in grain size of Mn-Zn ferrite from 11.38 to 67.42 nm and lattice parameter from 8.409 to 8.483 Å with rising sintering temperature. Gabal et al. [7] observed that the lattice constant of Mn-Zn decreased from 8.4466 to 8.4164 Å and density increased from 5.13 to 53.2 g/cm^3 with rising Zn concentration. Phong et al. [8] confirmed the cubic

M. Ikram et al., *Ferrite Nanomaterials for Wastewater Treatment*, Engineering Materials, https://doi.org/10.1007/978-3-032-09141-3_3

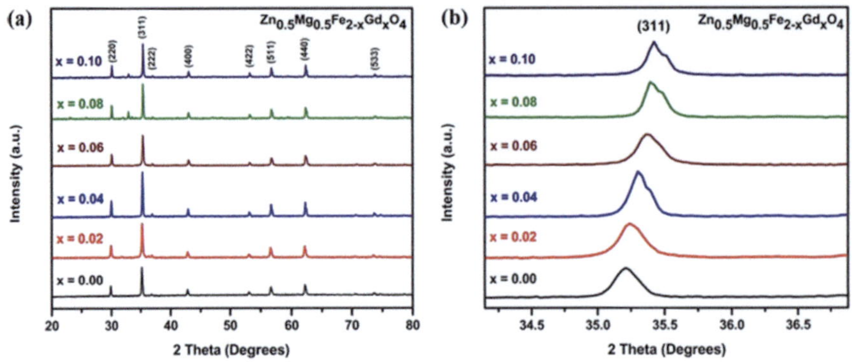

Fig. 3.1 **a** Diffraction patterns and **b** slight shifting of the intense peak of $Zn_{0.5}Mg_{0.5}Fe_{2-x}Gd_xO_4$. Reproduced with permission from [4]

phase (space group ~ Fd3m) of Mn-Zn ferrites with a lattice constant of 8.432 Å and density of 5.27 g/cm^3. Diffraction peaks of In-doped Mn-Zn, as examined by Kumar and colleagues [9], exhibited spinel phase (space group ~ Fd3m). With a higher amount of In, Bragg's peaks towards higher angles were observed with improved crystallite size from 14.6 to 15.9 nm and density from 5.315 to 5.445 g/cm^3. Vani et al. synthesized the cubic phase of $CoTb_yFe_{2-y}O_4$ and observed a reduction in grain size from 48 to 19 nm with increasing amount of Tb. Density was increased from 5.160 to 5.420 g/cm^3 and lattice parameter varied from 8.352 to 8.367 Å (Fig. 3.2) [10]. R. Srimathi et al. used a facile technique to prepare Al-doped $MgFe_2O_4$ and observed an enhancement in crystallite size from 34 to 46 nm with a higher amount of doping. The lattice constant varied from 8.365 to 8.346 Å, and the cell volume enhanced from 585.32 to 589.95 Å^3 [11]. Saha et al. used sol sol-gel approach to synthesize Ag-doped $NiFe_2O_4$, and the crystallite size was increased from 45 to 78 nm [12]. Kogulakrishna et al. prepared Zn-doped $MnFe_2O_4$, and a reduction in grain size was observed from 18.436 to 12.745 nm. X-ray density increased from 4.295 to 5.291 g/cm^3, and lattice constant varied from 8.934 to 8.446 Å [13].

3.2 Morphological Properties

To examine the morphology of FNMs, SEM, TEM, HRTM, and AFM approaches were utilized. SEM is the most often employed method, yet it offers inferior resolution in comparison to TEM. Dhiman et al. [14] confirmed the aggregated cubic-shaped NPs like morphology of $ZnFe_2O_4$. This aggregated behavior is observed due to the magnetic characteristics of FNMs. Uke et al. [15] synthesize the spherical NPS of Zn-MgFe2O4 with a size of 50–100 nm. Naidu et al. [16] examined the platelet-like particles of $Ni_{0.2}Mg_{(0.8-x)}Zn_xFe_2O_4$ and the enhancement in size with increasing concentration of Zn. TEM images revealed by Phong's observation [8]

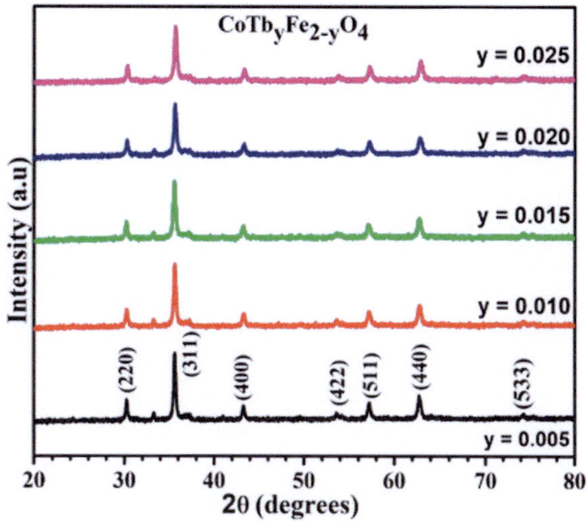

Fig. 3.2 XRD pattern of $CoTb_yFe_{2-y}O_4$. Reproduced with permission from [10]

indicated the homogenous spherical NPs morphology of FNMs. SEM images exhibited by Jalaiah's investigation [17] showed small grains on the FNMs surface, which increased with higher Ni content. In the TEM examination conducted by Angadi et al. [18], the particles exhibited little agglomeration resulting from the gradual development of particles throughout processing. The particle size of $Mn_{0.5}Zn_{0.5}Fe_2O_4$ was around 20 nm. The SEM images [19] revealed that pristine and modified MnZn ferrites exhibited a polyhedral shape with inhomogeneous particles. The grain size increased from 21.0 to 3.13 μm with increasing sintering temperature from 1150 to 1250 °C. The SEM examination of MnZn ferrites conducted by Yadav et al. [20] revealed that particle size increased with the addition of Sm content; however large-sized particles were mostly generated by the aggregation of ultrafine particles. The TEM scans indicated that all NPs were approximately spherical with a size of 10–20 nm. SEM examination [9] revealed uniformly spherical, lightly aggregated NPs. The morphology of MnZn ferrites is typically spherical, with a particle size ranging from 9 to 23 nm. Figure 3.3 displays the FE-SEM images of $CoTb_yFe_{2-y}O_4$ samples, with y ranging from 0 to 0.025 in increments of 0.005. The images demonstrated significant agglomeration of spherical grains, characterized by homogeneous appearance and non-uniform dispersion [10]. Ahmad et al. [21] synthesized $CoSm_xCe_yFe_{2-x-y}O_4$ and exhibited the dispersed grain distribution, with soft aggregates present owing to the abundance of rare earth ions (Ce and Sm) at the grain boundaries, which may impede grain formation. The mean grain size was determined to be 0.15 μm and reduced with increasing amount of substituted rare earth ions. Gingasu et al. studied cobalt ferrite and found micro-sized agglomerated NPs with a size ranging from 10 to 15 nm [22]. Aafiya et al. [23] prepared Zn-doped $NiFe_2O_4$ and demonstrated irregular shapes and dimensions of NPs. From SEM images, Kumar et al. [24] found the aggregated NPs of pristine $NiFe_2O_4$ with a size of around 23 nm. The examination of SEM images reveals alterations in the shape and dimensions of particles due to Y doping. The inclusion of bigger Y ions leads to an increase in particle size. The size determined using XRD measurement exhibits a similar pattern but varies in magnitude compared to the SEM observation. Kumar et al. [2] used the auto combustion method to synthesize $Mn_{1-x}Zn_xFe_2O_4$. The TEM micrographs indicate that the produced particles have a uniform spherical shape and are partially agglomerated. This aggregation is linked to Van der Waals forces and the single-domain nature of minute ferrite NPs, resulting in the permanent magnetization of NPs. SAED image clarified polycrystalline behavior of $Mn_{1-x}Zn_xFe_2O_4$ with circular rings (Fig. 3.4). Kogulakrishnan et al. exhibited the well-defined grains while also displaying the creation of agglomerates (irregular in shape and size) of Zn-doped $MnFe_2O_4$. Due to their magnetic characteristics, all materials exhibit a propensity to amalgamate into bigger aggregates. The histogram indicates average particle sizes of 59, 53, 36, 44, and 19 nm, respectively (Fig. 3.5) [13].

3.3 FTIR Spectroscopy

FTIR is the predominant analytical method in infrared spectroscopy. It provides estimations for the development of the spinel phases and analyzes the occurrence of different functional groups in the sample. It operates on the concept that as infrared (IR) radiation traverses a sample, a portion is absorbed, and another portion is transmitted by the sample. The emitted radiation gets recorded, resulting in a spectrum. The resultant spectrum generates a molecular identification of the material. Revathi et al. [25] characterize the FTIR spectra of Ni-doped $CoFe_2O_4$ in the range 4000–350 cm^{-1}. The bands at 1626 and 3424 cm^{-1} correspond to bending -OH and stretching -OH groups, respectively. The band from 450 to 385 cm^{-1} is linked to M-O modes of tetrahedral sites, and bands among 600–550 cm^{-1} ascribed to M-O modes at octahedral sites. With Ni doping, peaks moved towards higher wavenumber, confirming the interaction among dopant and $CoFe_2O_4$ (Fig. 3.6) [25]. Khishigdemberel et al. [26] prepared the Cu doped $MgFe_2O_4$ and characterized its transmittance properties. The most intriguing aspect of the IR spectra within the 800–400 cm^{-1} range was the bands corresponding to tetrahedral ($\nu_1 \sim 580$–569 cm^{-1}) and octahedral sites ($\nu_2 \sim 426$–408 cm^{-1}). Shobana et al. [27] reported a synthesis of Gd-doped $ZnFe_2O_4$ using a facile technique. The bands in the 600–400 cm^{-1} confirmed the synthesis of spinel ferrites. The bands at 522–534 and 414–457 cm^{-1} are tetrahedral and octahedral metal stretching modes, respectively.

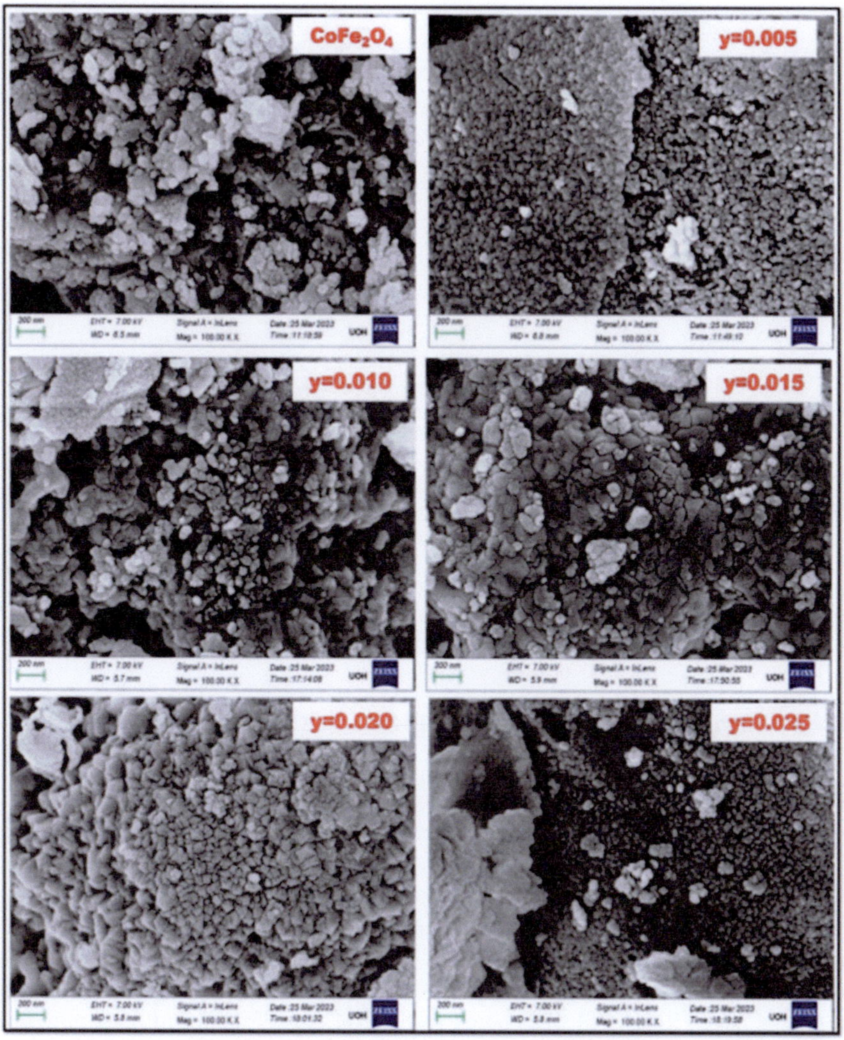

Fig. 3.3 SEM images of $CoTb_yFe_{2\text{-}y}O_4$. Reproduced with permission from [10]

3.4 Magnetic Properties

The predominant methods for assessing the magnetic characteristics of FNMs include vibrating sample magnetometer (VSM), magnetization hysteresis (M-H) loops, and electron spin resonance (ESR) hysteresis loop studies. These characterization approaches enable the calculation of saturation magnetization, remanent magnetization, and coercivity.

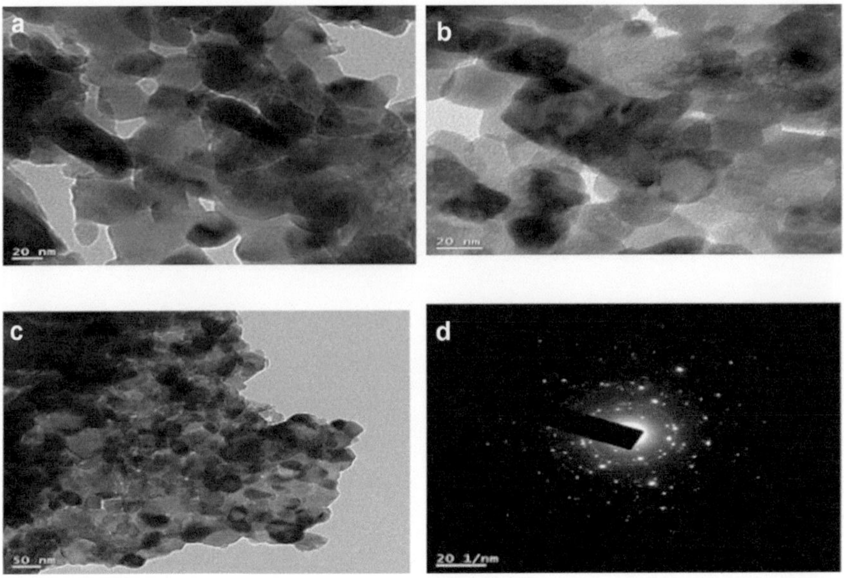

Fig. 3.4 **a–c** TEM images of $Mn_{1-x}Zn_xFe_2O_4$ ($X = 0.0, 0.6, 0.8$) and **d** SAED pattern of $Mn_{0.8}Zn_{0.2}Fe_2O_4$. Reproduced with permission from [2]

3.4.1 Saturation Magnetization (M_s)

Saturation magnetization refers to the maximum magnetization attained by a ferromagnetic material (Fe, Ni, Co). The interior of the magnetic material is typically partitioned into several domains; however, when the external MF intensifies, domain walls may shift and magnetization may realign within the domains, resulting in a single-domain state for the magnetic material. M_s is attained when the easy magnetization axis aligns with the direction of the MF, and the corresponding value of magnetization is referred to as M_s. The M_s of MnZn ferrites is elevated compared to other FNMs. Syue et al. [28] noted that the M_s rose from 11 to 62 emu/g with rising Mn^{2+} level and subsequently reached saturation. The M_s values ranged from 11.090 emu/g to 60.868 emu/g when the combustion process was employed without following thermal treatments. Praveena et al. [29] noted that the M_s rose from 53.33 to 78.26 Am^2/Kg with rising zinc content, then decreased to 10.74 Am^2/Kg with additional increments in Zn concentration. Prasad et al. [30] produced MFe_2O_4 (where M = Cu, Ni, Zn, Ni, and Co) by an auto-combustion technique. Among all synthesized spinel ferrites, Co-based FNMs exhibited the greatest M_s at 41 emu/g, whereas Zn-based FNMs had a much lower M_s of 12.87 emu/g (Fig. 3.7a). Mohamed et al. [31] documented the M_s and coercivity values of $ZnFe_2O_4$, which depend on the cobalt doping ratio. Substituting Zn^{2+} ions with Co^{2+} resulted in a rise in size from 8.3 to 11.4 nm, correlating with the elevated Co^{2+} doping concentration. The study demonstrated that the saturation magnetization value escalated with increasing Co^{2+}

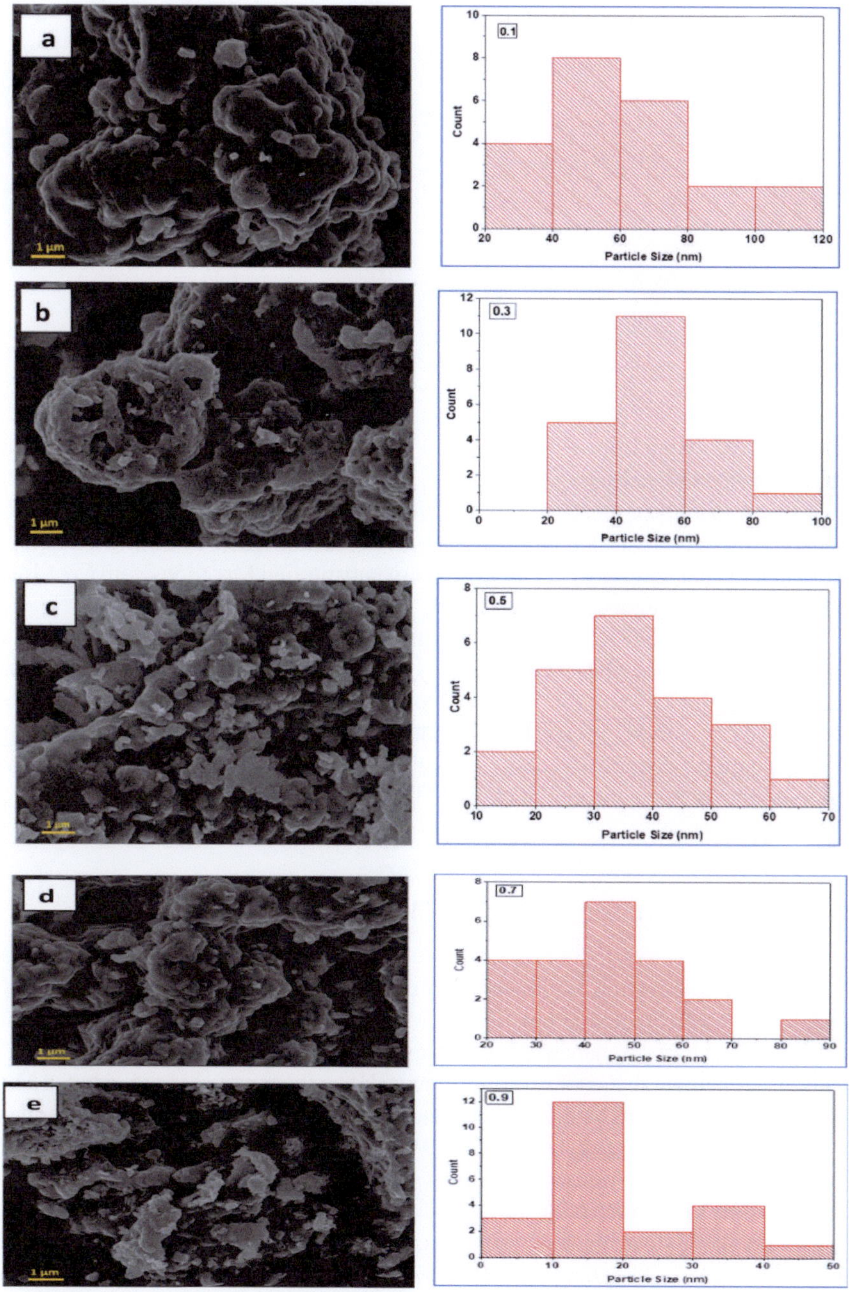

Fig. 3.5 FESEM images and histogram for $Mn_{1-x}Zn_xFe_2O_4$ of **a** 0.1, **b** 0.3, **c** 0.5, **d** 0.7 and **e** 0.9. Reproduced with permission from [13]

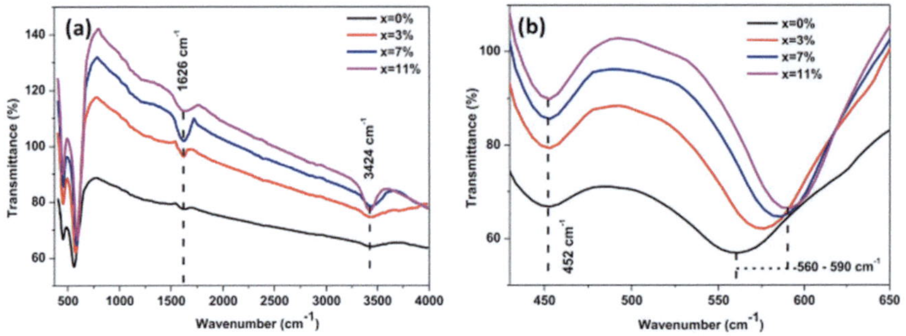

Fig. 3.6 FTIR spectra of pristine and Ni-$CoFe_2O_4$. Reproduced with permission from [25]

concentration, rising from 40 emu/g at x = 0.0 to 98 emu/g at x = 0.4. The magnetic anisotropy markedly altered when the Co^{2+} concentration in the $ZnFe_2O_4$ host lattice rose (Fig. 3.7b). Mazen et al. [32] studied the effects of Ni^{2+}, Cu^{2+}, and Co^{2+} replacement on the magnetic and structural features of $Li_{0.25}Mn_{0.5-x}M_xFe_{2.25}O_4$. The findings demonstrated the mutually beneficial effect of substituting Mn^{2+} ions with Co^{2+} ions in improving the magnetic properties of Li–Mn ferrites (Fig. 3.7c).

Purnama et al. [34] investigated the influence of calcination temperature on the magnetic features of $CoFe_2O_4$ produced by the co-precipitation method. It is evident that the migration of small Co^{2+} and Fe^{3+} ions among tetrahedral and octahedral sites results in a boost in M_s values. Furthermore, it was concluded that the enlargement of the magnetic domain size, which facilitates magnetization, accounted for the observed rise in magnetization with particle size (Fig. 3.7d).

3.4.2 Remanent Magnetization (M_r)

Remanent magnetization is the remaining magnetism that persists in a material after the magnetizing action has ceased. M_r is an essential property of a magnetic substance. The term "remanence" originates from the origin of the word "remanent," denoting something that endures or continues. Upon the removal of the induced MF, the term refers to the remaining magnetization that persists in the material. Dhiman et al. [35] noted a decrease in the M_r of $Mg_{0.9}Mn_{0.1}Sc_xFe_{2-x}O_4$ with increasing Sc concentration. The M_r value fell from 0.43 to 0.007 emu/g. The analysis revealed a steady decline in magnetization during the experiment, resulting in a reduction of M_r. Angadhi et al. [18] investigated the magnetic properties of $Mg_{0.5}Zn_{0.5}Sc_xFe_{2-x}O_4$ and noted a decrease in M_r values from 0.43 to 0.007 emu/g with increasing Sc concentration. The higher Sc concentration reduced the exchange interaction among the A and B sites, resulting in a reduction of the M_r. Das et al. [36] have observed the diminishing tendency of magnetic M_r in $CoY_xFe_{2-x}O_4$ ferrites. The paramagnetic characteristics of the Y ion are responsible for the reduction in M_S and M_r.

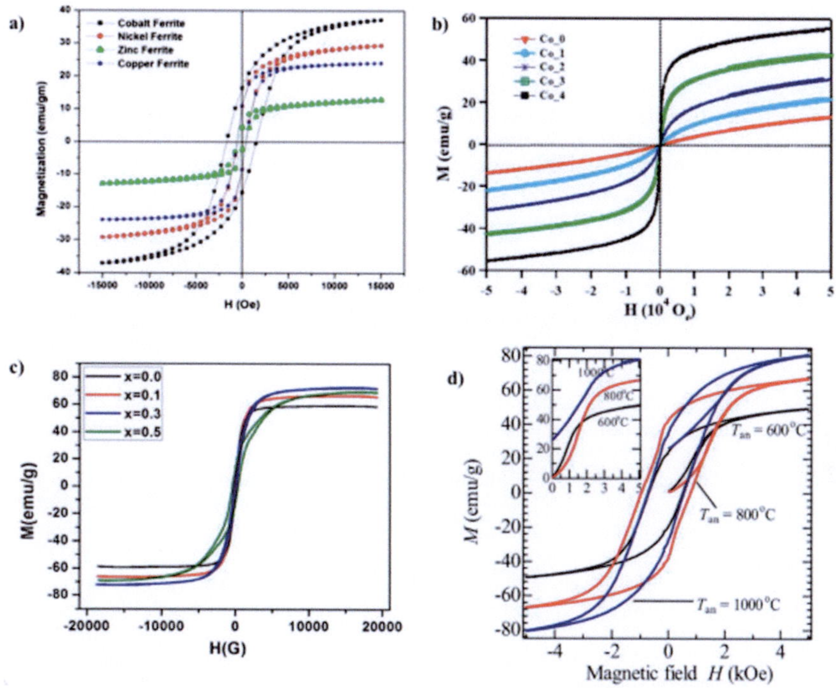

Fig. 3.7 Hysteresis curves. **a** MFe_2O_4, **b** $Co_xZn_{1-x}Fe_2O_4$, **c** $Li_{0.25}Mn_{0.5-x}Co_xFe_{2.25}O_4$ and **d** Co-precipitated $CoFe_2O_4$ NPs at different calcination temperatures (600, 800 and 1000 °C)**.** Reproduced with permission from [33]

3.4.3 Coercive Force (H_C)

To nullify the magnetization of a magnetic substance, a specific intensity of reversed MF must be exerted. Coercive force denotes the intensity of the external MF necessary to entirely eradicate a magnetic body's magnetization. The internal magnetic body is oriented in one direction, whereas the external MF is oriented in the other way. The coercivity of a material is sometimes referred to as magnetic coercivity, coercive field, or coercive force. Dhiman et al. [35] investigated the hysteresis curve to assess the coercivity of Sc-doped Mg-Mn ferrites. They found that the H_C decreased when the Sc concentration rose from 36.93 to 2.28 Oe. The researchers elucidated that the ferrites under investigation have soft magnetization, and the reduction in HC corroborates the soft characteristics of these samples when Sc^{3+} substitutes at the Fe^{3+} site. Angadhi et al. [18] observed that the coercive field first decreased in the sample with $x = 0.01$. The coercive field increased consistently from 58.92 to 84.32 Oe when the Sc concentration escalated from $x = 0.01$ to $x = 0.05$.

References

1. Dora, B.B., Kumar, S., Sahu, M.C.: Size controlled synthesis and magnetic behaviour of mg-zn nano ferrites by using aloe vera extract solution. Int. J. Pharm. Sci. Rev. Res **29**, 307–311 (2014)
2. Kumar, D.R., Lincoln, C.A., Ravinder, D., Ahmad, S.I.: Structural, morphological, luminescence, magnetic, and electrical transport properties of zinc-doped MnFe2O4 nanomaterials. Appl. Phys. A **126**(9), 705 (2020)
3. Singh, S.B., Srinivas, C., Tirupanyam, B.V., Prajapat, C.L., Singh, M.R., Meena, S.S., Bhatt, P., Yusuf, S.M., Sastry, D.L.: Structural, thermal and magnetic studies of MgxZn1− xFe2O4 nanoferrites: study of exchange interactions on magnetic anisotropy. Ceram. Int. **42**(16), 19179–19186 (2016)
4. Thakur, P., Kumari, S., Singh, S., Ravelo, B., Wan, F., Thakur, A.: Mg-Zn Ferrite nanoparticles: a brief review on synthesis, characterizations, and applications. ChemBioEng Rev. **10**(5), 711–736 (2023)
5. Ramiza, F.A., Ajmal, S.K., Khan, M.B., Nasim, A., Jamil, Y., Kashif, K., Amira, S.: Effect of UV radiations to control particle size of Mn-Zn spinel ferrite nano-particles. IOP Conf. Ser. Mater. Sci. Eng. **146**(1), 012029). IOP Publishing (2016)
6. Thakur, P., Sharma, R., Sharma, V., Sharma, P.: Structural and optical properties of Mn0. 5Zn0. 5Fe2O4 nano ferrites: effect of sintering temperature. Mater. Chem. Phys. **193**, 285–289 (2017)
7. Gabal, M.A., Al-Luhaibi, R.S., Al Angari, Y.M.: Mn–Zn nano-crystalline ferrites synthesized from spent Zn–C batteries using novel gelatin method. J. Hazard. Mater. **246**, 227–233 (2013)
8. Phong, P.T., Nam, P.H., Manh, D.H., Tung, D.K., Lee, I.J., Phuc, N.X.: Studies of the magnetic properties and specific absorption of Mn0. 3Zn0. 7Fe2O4 nanoparticles. J. Electron. Mater. **44**(1), 287–294 (2015)
9. Kumar, S., Shinde, T.J., Vasambekar, P.N.: Study of conduction phenomena in indium substituted Mn–Zn nano-ferrites. J. Magn. Magn. Mater. **379**, 179–185 (2015)
10. Vani, K., Hashim, M., Rana, G., Ismail, M.M., Batoo, K.M., Hadi, M., Kumar, N.P., Naveena, G., Sathish, B., Sriramulu, G., Devi, S., Ravinder, D.: Impact of rare earth Tb3+ substitution in cobalt ferrites: Tuning structural, dielectric, magnetic properties and photocatalytic activity. Ceram. Int. **51**(1), 240–251 (2025)
11. Srimathi, R., Rao, N.S., Merlin, A., Kiruthika, R., Selvaraj, A., Abdelkader, O.H., Dash, C.S., Revathi, S., Ahamed, A., Rajabathar, J.R., Sundararajan, M., Yuvaraj, S., Rajadurai, L.: Investigation of structural, magnetic, optical and dielectric characteristics of Al-doped MgFe2O4 nanoparticles. Solid State Sci. **159**, 107761 (2025)
12. Saha, S., Routray, K.L., Hota, P., Dash, B., Yoshimura, S., Ratha, S., Nayak, T.K.: Structural, magnetic and dielectric properties of green synthesized Ag doped NiFe2O4 spinel ferrite. J. Mol. Struct. **1302**, 137409 (2024)
13. Kogulakrishnan, K., Nithiyanantham, S., Mohan, R., Giridharan, N.V., Silambarasan, S., Venkadesh, A., Latha, V., Gunasekaran, B., Palaniappan, L.: Structural, electrical and magnetic studies on Zn doped MnFe2O4 nanoparticles with via Sol-Gel approach. J. Mol. Struct. **1321**, 140026 (2025)
14. Dhiman, P., Dhiman, N., Kumar, A., Sharma, G., Naushad, M., Ghfar, A.A.: Solar active nano-Zn1− xMgxFe2O4 as a magnetically separable sustainable photocatalyst for degradation of sulfadiazine antibiotic. J. Mol. Liq. **294**, 111574 (2019)
15. Uke, S.J., Mardikar, S.P., Bambole, D.R., Kumar, Y., Chaudhari, G.N.: Sol-gel citrate synthesized Zn doped MgFe2O4 nanocrystals: a promising supercapacitor electrode material. Mater. Sci. Energy Technol. **3**, 446–455 (2020)
16. Madhuri, W.: Microwave assisted solid state reaction method: investigations on electrical and magnetic properties NiMgZn ferrites. Mater. Chem. Phys. **181**, 432–443 (2016)
17. Jalaiah, K., Babu, K.V.: Structural, magnetic and electrical properties of nickel doped Mn-Zn spinel ferrite synthesized by sol-gel method. J. Magn. Magn. Mater. **423**, 275–280 (2017)
18. Angadi, V.J., Choudhury, L., Sadhana, K., Liu, H.L., Sandhya, R., Matteppanavar, S., Rudraswamy, B., Pattar, V., Anavekar, R.V., Praveena, K.: Structural, electrical and magnetic properties of Sc3+ doped Mn-Zn ferrite nanoparticles. J. Magn. Magn. Mater. **424**, 1–11 (2017)

19. Rahaman, M.D., Nahar, K.K., Khan, M.N.I., Hossain, A.A.: Synthesis, structural, and electromagnetic properties of Mn0. 5Zn0. 5− xMgxFe2O4 (x= 0.0, 0.1) polycrystalline ferrites. Phys. B **481**, 156–164 (2016)
20. Yadav, N., Kumar, A., Rana, P.S., Rana, D.S., Arora, M., Pant, R.P.: Finite size effect on Sm3+ doped Mn0. 5Zn0. 5SmxFe2− xO4 (0≤ x≤ 0.5) ferrite nanoparticles. Ceram. Int. **41**(7), 8623–8629 (2015)
21. Ahmad, S.I., Ansari, S.A., Kumar, D.R.: Structural, morphological, magnetic properties and cation distribution of Ce and Sm co-substituted nano crystalline cobalt ferrite. Mater. Chem. Phys. **208**, 248–257 (2018)
22. Gingasu, D., Mindru, I., Culita, D.C., Calderon-Moreno, J.M., Bartha, C., Greculeasa, S., Iacob, N., Preda, S., Oprea, O.: Structural, morphological and magnetic investigations on cobalt ferrite nanoparticles obtained through green synthesis routes. Appl. Phys. A **127**(12), 892 (2021)
23. Abushad, M., Arshad, M., Naseem, S., Ahmed, H., Ansari, A., Chakradhary, V.K., Husain, S., Khan, W.: Synthesis and role of structural disorder on the optical, magnetic and dielectric properties of Zn doped NiFe2O4 nanoferrites. J. Mol. Struct. **1253**, 132205 (2022)
24. Kumar, P., Rana, G., Dixit, G., Kumar, A., Sharma, V., Goyal, R., Sachdev, K., Annapoorni, S., Asokan, K.: Structural, electrical and magnetic properties of dilutely Y doped NiFe2O4 nanoparticles. J. Alloy. Compd. **685**, 492–497 (2016)
25. Revathi, J., Abel, M.J., Archana, V., Sumithra, T., Thiruneelakandan, R.: Synthesis and characterization of CoFe2O4 and Ni-doped CoFe2O4 nanoparticles by chemical Co-precipitation technique for photo-degradation of organic dyestuffs under direct sunlight. Phys. B **587**, 412136 (2020)
26. Khishigdemberel, I., Uyanga, E., Hirazawa, H., Sangaa, D.: Influence of Cu dope on the structural behavior of MgFe2O4 at various temperatures. Phys. B **544**, 73–78 (2018)
27. Shobana, M., Raguram, T., Eswaramoorthy, N., Cardenas-Jiron, G., Granados-Tavera, K., Alodhayb, A.N., Alzahrani, K.E., Pandiaraj, S., Geethapriya, J., Chethan Y.R.: Influence of Gd doping on the structural, optical, dielectric, and magnetic properties of ZnFe2O4: a combined experimental and theoretical study. Ceram. Int. **51**(5), 6748–6761 (2025)
28. Syue, M.R., Wei, F.J., Chou, C.S., Fu, C.M.: Magnetic and electrical properties of Mn–Zn ferrites synthesized by combustion method without subsequent heat treatments. J. Appl. Phys. **109**(7) (2011)
29. Praveena, K., Katlakunta, S., Virk, H.S.: Structural and magnetic properties of Mn-Zn ferrites synthesized by microwave-hydrothermal process. Solid State Phenom. **232**, 45–64 (2015)
30. Vara Prasad, B.B.V.S., Ramesh, K.V., Srinivas, A.: Structural and magnetic studies of nano-crystalline ferrites MFe2O4 (M= Zn, Ni, Cu, and Co) synthesized via citrate gel autocombustion method. J. Supercond. Novel Magn. **30**(12), 3523–3535 (2017)
31. Mohamed, W.S., Alzaid, M., SM Abdelbaky, M., Amghouz, Z., García-Granda, S., M. Abu-Dief, A.M.: Impact of Co2+ substitution on microstructure and magnetic properties of CoxZn1-xFe2O4 nanoparticles. Nanomaterials **9**(11), 1602 (2019)
32. Mazen, S.A., Elsayed, H.M., Abu-Elsaad, N.I.: Effect of divalent metal ions substitution on structural and magnetic properties of Li0. 25Mn0. 5-xMxFe2. 25O4 (M= Co2+, Ni2+, Cu2+) spinel ferrites. Mater. Chem. Phys. **256**, 123676 (2020)
33. Salih, S.J., Mahmood, W.M.: Review on magnetic spinel ferrite (MFe2O4) nanoparticles: from synthesis to application. Heliyon **9**(6), e16601 (2023)
34. Purnama, B., Wijayanta, A.T.: Effect of calcination temperature on structural and magnetic properties in cobalt ferrite nano particles. J. King Saud Univ.-Sci. **31**(4), 956–960 (2019)
35. Dhiman, M., Rana, S., Singh, M., Sharma, J.K.: Magnetic studies of mixed Mg–Mn ferrite suitable for biomedical applications. Integr. Ferroelectr. **202**(1), 29–38 (2019)
36. Das, S.B., Singh, R.K., Kumar, V., Kumar, N., Singh, P., Naik, N.K.: Structural, magnetic, optical and ferroelectric properties of Y3+ substituted cobalt ferrite nanomaterials prepared by a cost-effective sol-gel route. Mater. Sci. Semicond. Process. **145**, 106632 (2022)

Chapter 4
Chemical Composition of Organic Dyes

Dyes are synthetic organic chemicals that are either water or oil soluble; however, pigments are insoluble and exist in particulate form [1]. Organic dyes are utilized in several industrial applications, including textile (the most prominent), food items, cosmetics, and paper printing, resulting in their common presence in industrial effluent [2]. The most prevalent dyes are azo dyes, as they are more facile and economically viable to synthesize compared to natural dyes. In 2003, the yearly output of dyes exceeded 7×10^5 tons [3]. Dyes are deemed environmentally dangerous due to their toxicity to living species, either directly or by their absorption and reflection of sunlight, which disrupts the development of aquatic life. Moreover, they provide color to effluent, resulting in aesthetic concerns [4, 5]. Dyes are intricate unsaturated chemical compounds that absorb light and impart color to the visible spectrum. A dye is a material that possesses the ability to absorb portions of the visible spectrum (chromophore) [6]. The color is determined by the percentage of light that isn't captured by the dye but reflected.

The conjugate double bonds in dyes provide a chemical structure suitable to light absorption.

Consequently, colorants frequently comprise aromatic amines. Dyes are classified by their ability to absorb sunlight within the visible spectrum [7]. The conversion of white light into colorful light by reflection on a surface or diffusion is due to selectively absorbing of energy by certain groups. The varying intensity of colors in different substances is associated with their chemical composition. A dye is a substance that absorbs specific wavelengths of light and then reflects the complementary coloration. The matrix, auxochrome, and chromophore are three fundamentals group present in structure of dyes. The chromophore serves as the active location of the dye, encapsulating the spatial arrangement of atoms that absorb light energy.

Molecules containing chromophores are widely utilized in surgical process, with synthetic dyes being more favored over natural optical adjuncts [8]. Dyes can be categorized into non-ionic, anionic (direct, acid and reactive dyes), and cationic (basic dyes) types [9].

M. Ikram et al., *Ferrite Nanomaterials for Wastewater Treatment*, Engineering Materials, https://doi.org/10.1007/978-3-032-09141-3_4

4.1 Cationic Dyes

4.1.1 Rhodamine B (RhB)

Rhodamine B (RhB) is a xanthene-derived dye classified under the basic group, exhibiting high solubility in water. The RhB-based materials exhibit intriguing physicochemical and photophysical capabilities attributable to their distinctive structural characteristics (Fig. 4.1a). Figure 4.1a demonstrates that both RhB and its derivatives include xanthene units. Consequently, while RhB is a renowned leuco dye, it belongs to the xanthene family. Figure 4.1b shows RhB in equilibrium between ring-closed and ring-open forms. The equilibrium changes to the left with elevated pH and to the right at reduced pH. In the previous instance (ring-closed), the material exhibits no absorption (in the visible spectrum) or emission owing to the existence of a spiro-sp^3 carbon atom. The spiro-carbon atom reduces the system's aromaticity by breaking the xanthene moiety's conjugation. In the latter scenario (ring-open), the material exhibits entirely contrasting behavior, demonstrating exceptional absorption and emission properties. The spiro-sp^3 carbon transitions to a planar sp^2 carbon atom, resulting in prolonged conjugation and heightened coloration of the system [10–12].

The RhB (closed-form) comprises the spirolactone component and xanthene moiety, which are oriented in planes that are orthogonal to one another (Fig. 4.2a). This leads to inadequate overlap of the π orbitals of each moiety, resulting in an excessively large HOMO–LUMO gap. Consequently, the closed structure fails to absorb

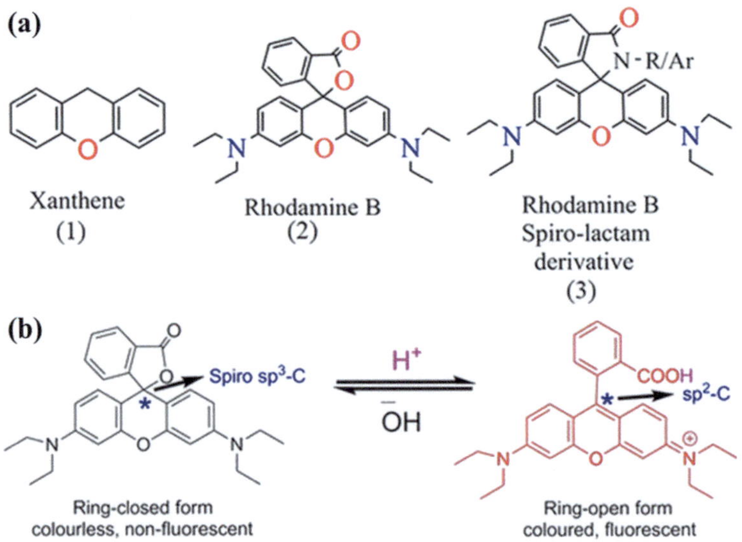

Fig. 4.1 **a** Illustration of a xanthene (1), RhB (2), and RhB spirolactam derivative (3). **b** RhB ring closed and open forms. Reproduced with permission from [10]

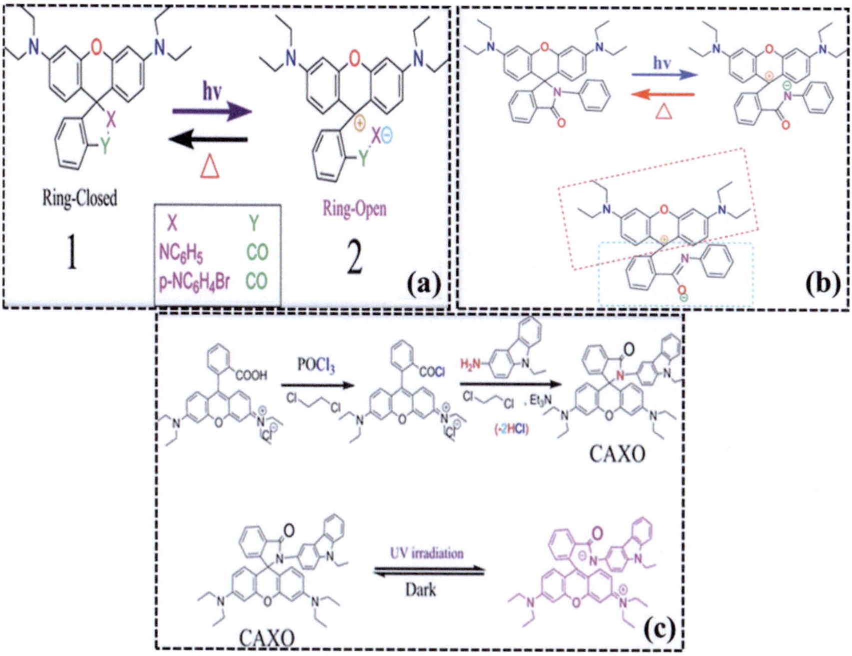

Fig. 4.2 **a** Photo-induced RhB derivative reaction among (1) colorless and (2) colored forms, **b** photochromic reaction of *N*-phenyl-rhodamine lactam, **c** synthesis of RhB derivative and its photochromism. Reproduced with permission from [10]

visible light and appears colorless. Conversely, in the open configuration of RhB, the spirolactone segment and xanthene moiety are almost co-planar. Consequently, longer conjugation transpires, resulting in a reduced energy gap between HOMO and LUMO. Consequently, a pronounced absorption is observed in the visible spectrum, indicating coloration [10].

For the first time, RhB derivatives were utilized as optical storage when researchers announced that they were photochromic in 1977 [13]. They demonstrated that exceptional reversibility occurs between forms 1 and 2 after photoisomerization of the synthesized material. In 1989, ten years later, Gleiter and his colleagues enhanced the notion of photochromism [14]. They investigated the mechanistic pathway of the phenomenon via kinetics for an initial time (Fig. 4.2b) and examined all aspects influencing the kinetics and quantum effectiveness. Consequently, Feng and his colleagues announced a photo-responsive material known as CAXO (carbazole xanthenes one) (Fig. 4.2c). The material facilitates a unique interaction among the xanthene (RhB component) as the electron receiver and the carbazole component as the electron donor, resulting in exceptional color-change performance under UV stimulation [15].

4.1.2 Methylene Blue

Mb ($C_{16}H_{18}N_3ClS$), an aromatic heterocyclic cationic dye (molecular weight ~ 319.85 g mol^{-1}), is a primary thiazine dye exhibiting a λ_{max} of 663 nm (Fig. 4.3). It is extremely soluble in water, hence creating a stable solution at ambient temperature [16–18]. The MB structure with its various resonance structures is shown in Fig. 4.3a and b.

MB (13.82 Å length, 9.5 Å width), is a dense, odorless, dark green powder at ambient temperature that produces a blue solution upon dissolution in water (solubility in water ~ 43.6 g/L). MB exhibits a distinctive dark blue color in its oxidized state and is devoid of color in its reduced form (leucoMB-LMB), shown in Fig. 4.4. The Auxochrome and chromophore determine the color of MB dye. The N-S conjugated system (chromophore) within the core aromatic, whereas the N-composed groups (auxochrome) possess a lone pair e^- on the benzene ring. Absorption spectra of MB include a prominent absorption peak at around 664 nm (shoulder around 612 nm linked to MB dimer) corresponding to monomer. Two extra bands emerge in the UV spectrum, with maxima at around 245 and 292 nm, which are linked to substituted benzene rings. The absorption intensity progressively diminishes as the photocatalytic process advances [19–22].

MB is an appealing molecule with diverse features beneficial for medical purposes and serves as an efficacious antibiotic for the treatment of anemia and malaria [23]. MB has predominantly been utilized in both human and animal medicine for various medical and therapeutic applications. It possesses several potential uses in the fabric, drugs, paper, printing, medical, and food sectors [24, 25]. It is the predominant dye in the textile sector and is regarded as one of the most favored colorants for garments. Textile industries often discharge significant quantities of MB dyes into environmental water bodies, posing a health risk to humans and microorganisms [26]. MB dye poses a health risk to humans at elevated concentrations because to its significant toxicity [24]. MB is poisonous, carcinogenic, non-biodegradable, and poses

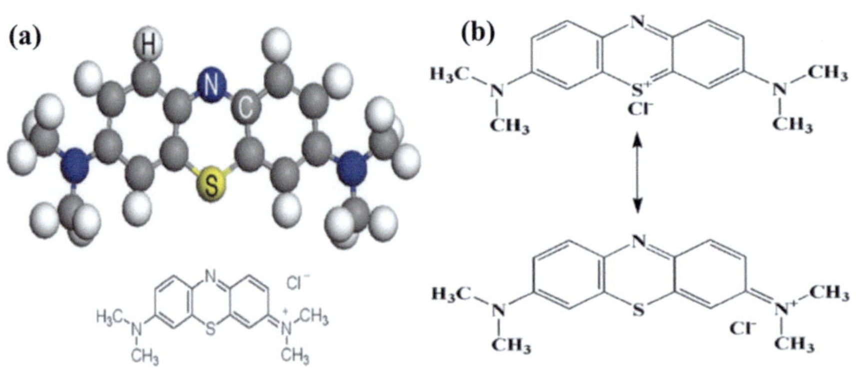

Fig. 4.3 **a** Structure of MB and **b** various resonances of MB. Reproduced with permission from [22]

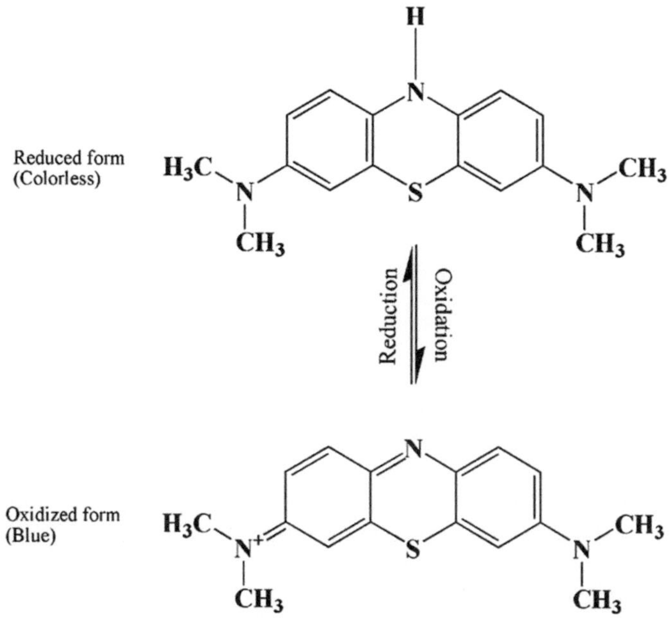

Fig. 4.4 Oxidation and reduction of MB dye. Reproduced with permission from [22]

a significant risk to human health as well as detrimental impacts on the environment. MB poses several health concerns to humans, including breathing problems, gastrointestinal issues, eyesight, and both digestive and mental diseases [27].

The release of MB into the ecosystem poses a considerable danger due to aesthetic and ecological concerns. It diminishes light penetration and serves as a hazardous source within ecosystems for organisms [28]. The presence of MB in aquatic environments, even at minimal concentrations, results in the formation of highly pigmented by-products. Due to its elevated molar absorption coefficient (~8.4×10^4 L mol^{-1} cm^{-1} at 664 nm), which diminishes sunlight transmission, it lowers the solubility of oxygen, impacts the photosynthetic activity of aquatic organisms, and diminishes the variety and visual appeal of the biological community [22].

4.1.3 Methyl Violet

Methyl violet (MV), sometimes referred to as crystal violet, is a cationic dye from the triphenylmethane group ($C_{25}H_{30}ClN_3$), commonly utilized for dyeing silk, wool, and cotton in the fabric sector (Fig. 4.5). MV has environmental persistence due to its stable molecular composition, characterized by $-NH_2$ groups and aromatic chains, resulting in a complex structure and extended half-life and it decomposes into toxic gases of CO_2, CO, and NO during burning or oxidation. The dye, in its raw state,

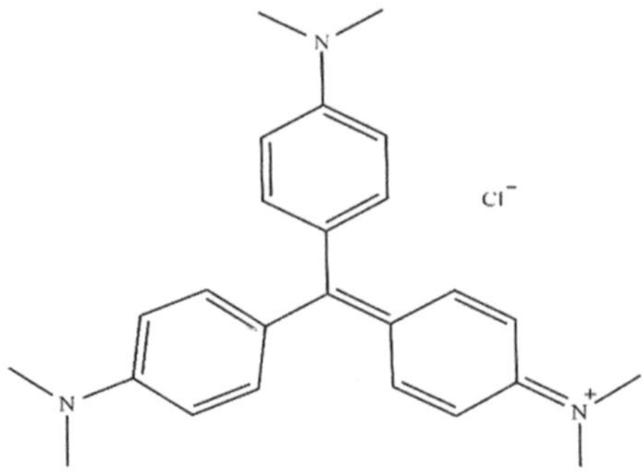

Fig. 4.5 Structure of methyl violet

is poisonous and carcinogenic, capable of entering the human body by inhalation or consumption, resulting in inflammation of skin and eyes, rapid heartbeat nausea, and shock. MV enters cytoplasm through the interaction with −ve cell membrane surface and maintains itself there. It may result in severe blindness, pulmonary and renal failure, and cancer. MV is a luminous water-soluble dye; hence, its existence in watery environments, even at concentrations below 1 mg/L, can impair the metabolic processes of aquatic plants [29].

4.1.4 Malachite Green

Malachite green (MG), a triarylmethane dye ($C_{23}H_{25}N_2Cl$, 364.91 g/mol), containing green crystals, is synthesized by combining benzaldehyde with dimethylaniline in the presence of strong H_2SO_4 or zinc chloride. Leuco-malachite green (LMG, $C_{23}H_{26}N_2$), containing brown crystals having a molecular weight of 330.46 g/mol, is the reduced form of MG (Fig. 4.6). From the 1930s, MG has become recognized for its fungicidal effects. MG was extensively employed as a localized fungicide and ecto-parasiticide within the international aquaculture sector [30, 31]. MG was demonstrated to be the most efficacious treatment for parasitic ectoparasites in the 1960s. The utilization of MG in fishing is highly esteemed, provided that the harmful effects and associated hazards to soil, aquatic species, and users of affected fish are acknowledged [32, 33]. MG is efficiently absorbed by fish and various other animal tissues, whereby it is transformed into its colorless derivative, LMG, which is lipophilic and persists in fatty tissues for extended durations, so entering the human food chain and accumulating in tissues [34, 35]. The removal of LMG is significantly affected by fat quantity and is excreted gradually (t1/2 40 days) in certain fatty fish tissues owing to its lipophilic properties [31]. Data indicates the presence of residual MG and its modified form (LMG) in the blood, kidneys, liver, muscles, and several other tissues, as well as in fried foods and eggs. MG causes sinusoidal obstruction and extensive destruction in the liver, along with mitochondrial impairment and alterations in hematological parameters. In rainbow trout, reduced hemoglobin and anemic responses

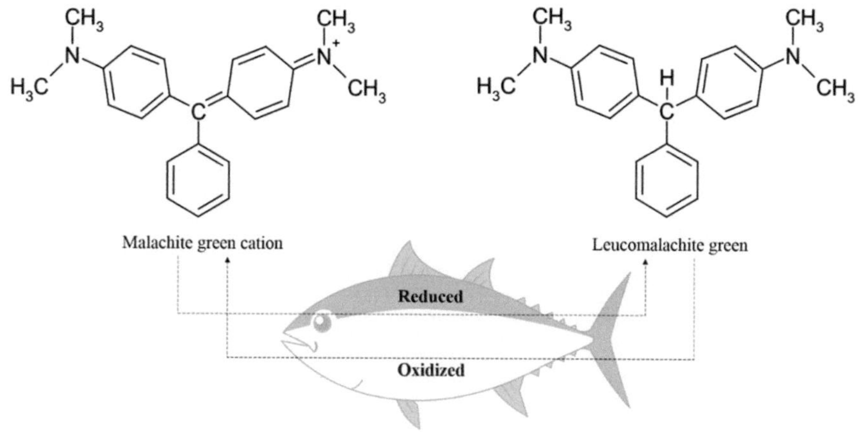

Fig. 4.6 Reduction of MG into LMG. Reproduced with permission from [31]

have been documented, leading to reproductive anomalies in the species [32, 35]. MG was recently identified as a tumor stimulant; nonetheless, our understanding of its carcinogenic potential remains restricted. LMG has fewer cancerous and mutagenic properties compared to MG, indicating that LMG does not significantly impact the lifespan of cells or DNA. The harmful effect of this compound induces genetic mutations, cancer development, pulmonary toxicity, and chromosome fractures, which escalate with temperature, exposure duration, and quantity, noting that 0.1 mg L^{-1} of MG can be fatal to fish. LMG has considerable stability during fish preparation, even at elevated temperatures, but the potential oxidation of LMG throughout the refrigeration or storage of fish tissues gradually reverts it to MG [31]. Nevertheless, several studies have indicated that heating fish muscle by methods such as roasting, boiling, or microwaving, as well as refrigerating or freezing, results in a reduction of MG and LMG residues in the muscle [30].

4.1.5 *Crystal Violet*

The compound known as crystal violet (CV, $C_{25}H3_{0}N3Cl$), which is composed of N, N, N1, N1, N11, and N11-hexa-methyl-para-rosaniline, is also a triphenylmethane dye that has been widely utilized in the field of medical treatment for humans and animals as a biological stain (Fig. 4.7). Additionally, it has been utilized as a textile dye in the business of textile processing. It is cationic in nature, known as gentian violet, and has one dimethyl-amino group on each phenyl ring. It is frequently employed as a purple dye for fabrics including cotton and silk and imparts an intense violet hue to paintings and printing ink [36–38].

Additionally, CV is utilized in the therapeutic industry as a mutagenic and antibacterial agent, with the purpose of preventing the growth of fungi in poultry feed. It

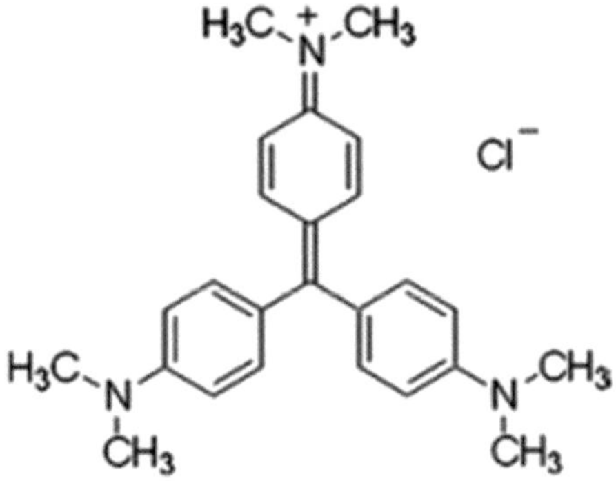

Fig. 4.7 Chemical structure of crystal violet

is identified as a persistent, resistant dye molecule that remains in the ecosystem for an extended duration and exerts harmful effects on both aquatic and terrestrial organisms. In-vitro studies have demonstrated that CV functions as a mitotic toxin, a powerful carcinogen, a strong clastogen, and promotes tumor formation in certain types of fish. Consequently, CV is classified as a biohazardous material. The dye is known to induce mild ocular irritation, severe photophobia, and irreversible damage to the retina, as it contains a cationic dye that is very poisonous to mammalian cells [39]. The dark-colored effluent from various industries, containing CV, adversely impacts the metabolic processes of aquatic plants owing to less exposure to sunlight and may also be hazardous to other aquatic organisms because of its high levels of aromatics and chlorides. The thin layer of dyes that accumulates on the surface of a collecting water body eventually diminishes metabolic processes of aquatic plants and decreases dissolved oxygen levels, while in agricultural soil, it obstructs seed germination and impedes crop development [38, 40]. Recently, there has been a heightened interest in the ecological regulation of dyes due to their toxicity and genotoxic impacts on living beings, as they include recognized carcinogens, including benzidine and other aromatic chemicals. Owing to its detrimental impact on human health, CV has been classified as a hazardous chemical or substance, resulting in its prohibition in fisheries and the food sector. Nonetheless, it is utilized in some regions because of its comparatively low expense, accessibility, and effectiveness. Dye sectors release around 10–20% of dyes in a free state with effluent throughout the dying process; however, this percentage may escalate to 50% when azo dyes are employed. The release of this substantial volume of dye with the introduction of wastewater into aquatic ecosystems may significantly impact aesthetic qualities and water clarity [38].

4.1.6 Basic Fuchsin

A triaminotriphenylmethane dye, known as Basic fuchsin ($C_{20}H_{20}ClN_3$), comprises a combination of three dyes: Magenta II, Pararosaniline, and Rosaniline, so referred

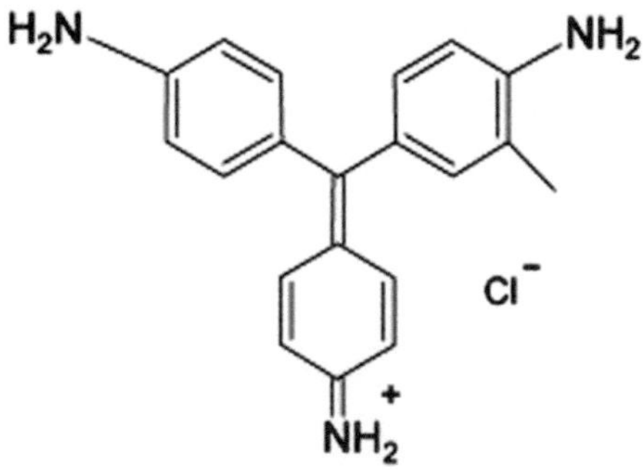

Fig. 4.8 Chemical structure of basic fuchsin dye. Reproduced with permission from [46]

to as Magenta II (Fig. 4.8) [41]. It is a remarkable dye that is inherently inflammable and utilized as a pigment for fabric and leather-based products. It has anesthetic, antibacterial (against gram-positive microbes), and fungicidal effects [42, 43]. Being in direct contact with the Basic fuchsin dye may result in significant irritation to the skin and eyes [44]. Consumption may result in intestinal itchiness, manifesting as feeling sick, throwing up, and having diarrhea, while inhaling of the dye can irritate the respiratory system. Long-term exposure to the dye may impact the neurological system, resulting in dizziness, headaches, lethargy, and muscular spasms. In humans and animals, its toxic effect encompasses cancerous and mutagenesis effects [45, 46].

4.1.7 Brilliant Green

Brilliant Green (BG) dye is an odorless a yellowish- to green powder utilized for several applications, including biological staining, dermatological treatments, animal health care, and as an ingredient in chicken feed to prevent the proliferation of the growth of mold, intestinal parasites, and fungi. It is also widely utilized in dyeing fabrics and printing on paper (Fig. 4.9) [47, 48]. Approximately 0.8–1.0 kg of BG is utilized for each ton of manufactured paper. In human beings, BG induces intestinal itching, with symptoms such as feeling sick, vomiting, and constipation [9]. It also induces inflammation of the lungs, resulting in cough and dyspnea. Dermal contact induces irritation characterized by erythema and discomfort. Brilliant Green can produce harmful byproducts such as CO_2, NO_x, and SO_x during thermal breakdown [49].

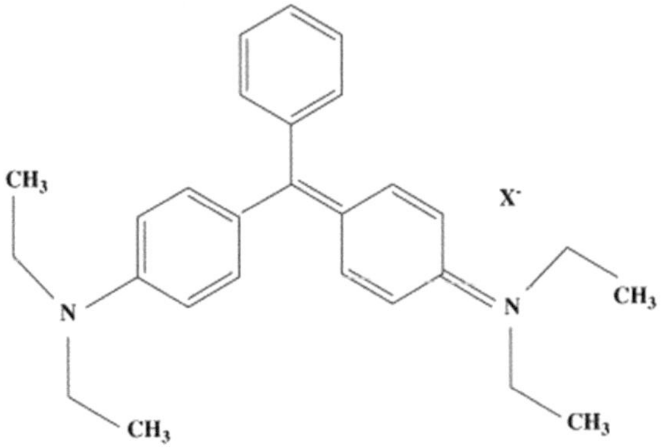

Fig. 4.9 Chemical structure of brilliant green dye. Reproduced with permission from [49]

4.2 Anionic Dyes

4.2.1 Methyl Orange

Methyl orange (MO, $C_{14}H_{14}N_3NaO_3S$, density ~ 1.28 g/cm^3, melting point > 300 °C) is an anionic azo dye characterized by good chemical durability and classified under the sulphonated azo group (Fig. 4.10). MO has partial solubility in ambient water, considerable solubility in heated water, and is insoluble in ethanol. It is weak acid, exhibiting a pH of around 6.5 in water, serves as pH indicator, red color in acidic condition and orange in basic environment [50]. The transitional period ranges from 3.0 to 4.4. MO is widely utilized in the printing, leather goods, fabric, and paper sectors. MO is extensively utilized in the textile sector and constitutes a significant portion of industrial waste effluent, contaminating several water bodies. The quantity of MO dye in textile wastewater can attain levels of up to 500 ppm. MO dye has mutagenic, cancer-causing, and harmful effects on aquatic life. It is a widely utilized, harmful anionic azo dye, characterized by its strong recalcitrance and resistance as a xenobiotic, imposing a considerable load on the environment. Azo dyes have persistence in the natural surroundings, and their biodegradation faces significant challenges [51, 52]. MO is a more detrimental dye that induces significant water contamination when discharged into the ecosystem. Its elevated levels in drinking water result in anemia, stomach discomfort, headaches, dizziness, cognitive disorientation, increased sweating, and nausea. MO inadvertently enters the body by eating, where intestinal microbes convert it into aromatic amines, which may induce cancer of the intestines. Consequently, effluent containing MO must be reduced and detoxified prior to environmental disposal [53].

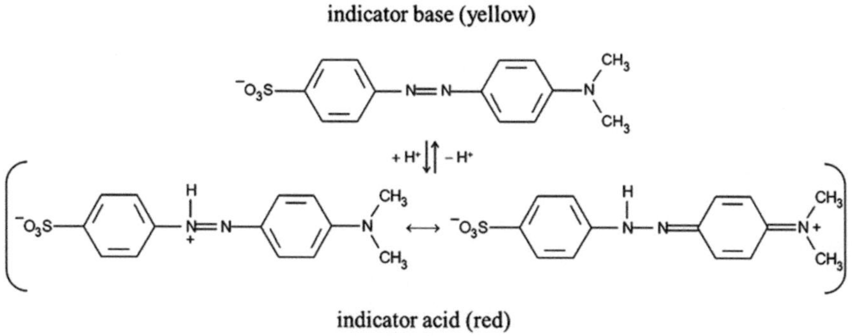

Fig. 4.10 MO structure (only one resonance structure is given for the indicator base and two for the indicator acid). Reproduced with permission from [54]

4.2.2 Congo Red

Congo-red (CR, $C_{32}H_{22}N_6Na_2O_6S_2$, molecular weight ~ 696.68 g mol^{-1}) is an azo dye identified by Paul Bottinger in 1884. CR dye, an acidic diazo dye, comprises (-N=N-) chromophores acidic auxochromes (-SO_3H) associated with benzene structures (Fig. 4.11) [55].

CR imparts a distinctive hue in an aqueous solution owing to the existence of an azo structure. It exhibits a red hue in a basic condition and a blue hue in an acidic environment. CR can generate an amine compound, including benzidine, upon the splitting of its azo groups. Benzidine is a prevalent carcinogen; hence, CR is classified as a prohibited azo dye [57]. Despite its beneficial uses across several sectors such as fabrics, pigments, leather goods, food, and paper, the widespread utilization of CR contributes to industrial pollution. CR negatively impacts the health of humans, resulting in many potentially deadly disorders when present in high concentrations due to its cytotoxicity, tumor-causing, and mutagenic properties. It impacts the ocular, integumentary, breathing, and genital systems. Benzidine, bladder carcinogen, along with its poisonous metabolite, induces allergic reactions and is a dangerous substance. Benzidine-based dyes induce hepatocarcinoma, nuclear

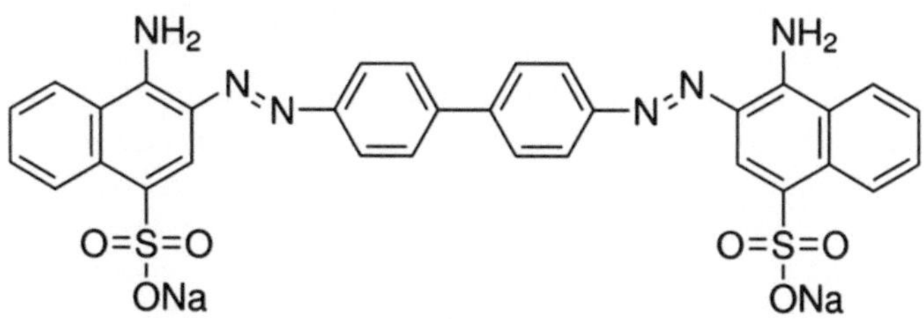

Fig. 4.11 Chemical structure of congo red dye. Reproduced with permission from [56]

anomalies, and chromosomal aberrations in mammalian cells during animal tests [56].

4.2.3 Eriochrome Black T

Eriochrome Black T (EBT, $C_{20}H_{12}N_3NaO_7S$) is an anionic azo dye predominantly utilized in the dyeing of fabrics, such as cotton fibers, wool, and silk [58]. This pigment has been employed as an indicator in titration and, more recently, in association with the textile sector and research institutions [59]. EBT belongs to the azo dye category, which constitutes 70 percent of all dyes produced economically and employed in the chemical sector. Furthermore, EBT is employed as an indicator in complexometric titrations for the purpose of detecting Zn^{2+}, Ca^{2+}, Ba^{2+}, and ions, in addition to also being employed for biological imaging [60]. In addition to its chemical stability, EBT has been categorized as hazardous dye. EBT is detrimental to aquatic creatures and may have adverse effects on aquariums over the long term or perhaps permanently. On account of the presence of sulfonate groups within the structure of EBT, there is a possibility that it poses a threat to both the aquatic environment and the human population [61]. The dye is hazardous, and the metabolites that are produced when it is broken down, such as naphthoquinone, are far more hazardous. Considering this, EBT is famously challenging to separate from aqueous solutions owing to its chemical configuration. When EBT is employed in the processing of contaminated wastewater with certain chemicals, it generates more complex chemical frameworks, hence enhancing its durability against photodegradation. The extraction of EBT from water is famously problematic due to its high water solubility and chemical durability [62].

4.2.4 Rose Bengal

Rose Bengal (RB), often referred to as Acid Red 94, is a fluorescein pink dye characterized by the presence of halogens (Fig. 4.12). It serves several medical diagnostic purposes, such as liver function assessments, staining of decaying ocular tissue, and devitalized retinal cells in conjunctivitis. RB is utilized in the treatment of both psoriasis and eczema at a minimal dosage. RB exhibits significant harmful impacts on human corneal epithelium, alongside its numerous therapeutic uses. The dye induces irritation, itching, desquamation, erythema, and occasionally vesication upon skin contact. Symptoms include inflammatory processes, erythema, lacrimation, and pruritus of the eyes. Upon inhalation, it obliterates mucous membranes, resulting in breathing difficulties in individuals [63]. It has a strong potential for intersystem crossings, which results in the production of a photoactive triplet excited state, and possesses remarkable optical and photochemical features, such as a substantial absorption coefficient in the visible range. The dye has been utilized in the process

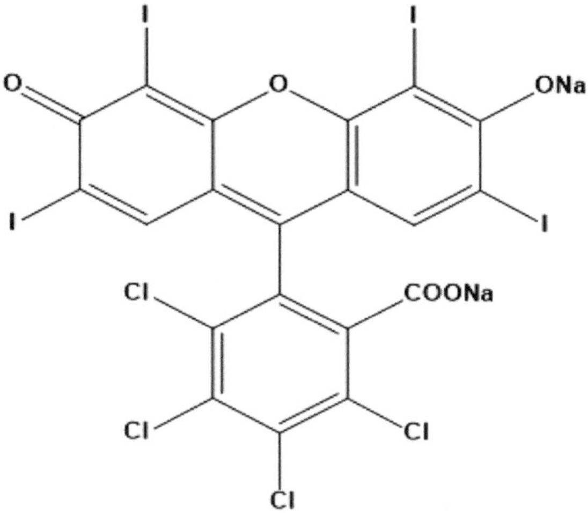

Fig. 4.12 Chemical structure of rose bengal dye. Reproduced with permission from [63]

of photodynamic suppression of catalase, photoinactivation of olefins through the production of singlet oxygen ($^1O_2$), and as a photosensitizing agent for the purpose of eliminating biological organisms including the E. coli microbes and vaccinia virus. It obstructs the development of bacterial populations and retards the dissemination of mold spores within the surroundings. Furthermore, it is employed to evaluate the nature of nucleotide binding sites in certain enzymes, photoinactivation, and to identify residues of amino acids in the binding sites of specific enzymes [64, 65].

4.2.5 *Eosin Y*

Eosin Y is a heterocyclic red color dye that contains bromine atoms. It is utilized in a variety of industries, including the printing process, coloring, leather goods, fluorescent pigment, and other related materials. As a result of its dark hue and poisonous nature, the immediate discharge of effluent that contains Eosin Y will make a significant contribution to the worldwide ecological crisis. Eosin Y is frequently employed as a counterstain in biological processes [66, 67]. Hematoxylin–eosin identification is predominantly utilized for histology sample staining. The nucleus will exhibit a blue to dark violet hue due to hematoxylin, while the eosin Y solution will impart a red–orange coloration to the cytoplasmic and interfacial substances. Nearly all published analytical methods employing the eosin reaction rely on the creation of an interaction complex among the dye and the analyte, with rare exceptions. The process of reaction was conducted mostly in a somewhat acidic environment. Consequently, the process of complex formation will be examined in relation to these media. Based on the solution's pH, eosin can manifest in many forms (Fig. 4.13), indicated by the equation:

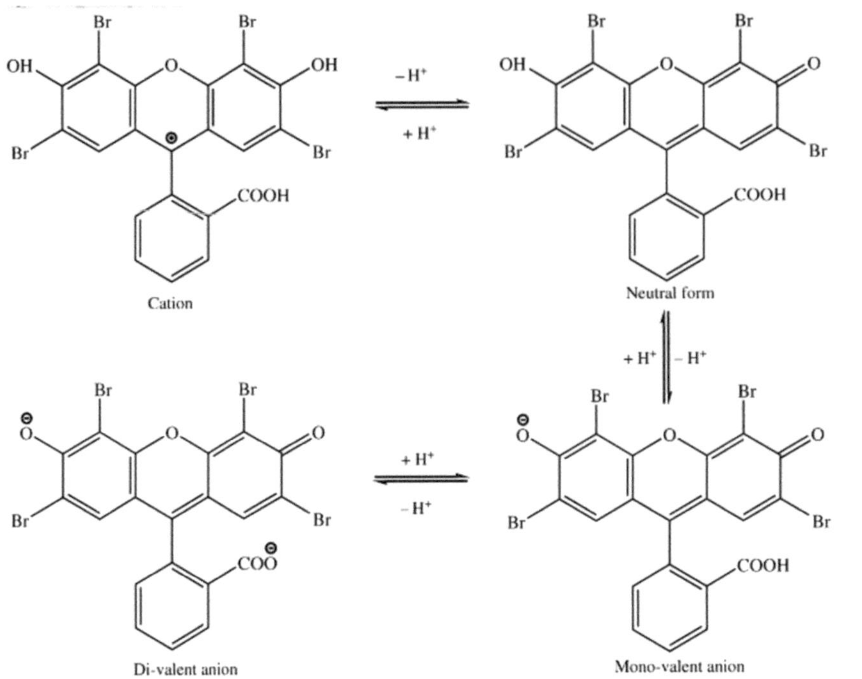

Fig. 4.13 Variations in eosin Y structure depending on pH of the medium. Reproduced with permission from [68]

$$H_3R^+ \rightleftarrows H2R \rightleftarrows HR- \rightleftarrows R^{2-}[68]$$

where R denotes the anionic component of dye. The values for pK_{a1}(2.10), pK_{a2}(2.85), and pK_{a3} (4.95) of eosin Y were reported in the literature. In a weakly acidic medium (pH 3.0–4.5), eosin Y mostly resides in the monovalent anionic state (HR^-). There are two potential mechanisms for the generation of the monovalent anion of eosin: ionization of the -OH group or the -COOH group. The ionization process of the -COOH group happened initially in fluorescein, followed by -OH group. In the instance of eosin Y, the proximity of two potent electron-withdrawing bromine atoms to the -OH group diminishes the charge density at the oxygen atom of the hydroxyl moiety. Consequently, the -OH exhibits a greater propensity for dissociation compared to the -COOH group. Consequently, the eosin monovalent anion is generated by the ionization of the -OH group. The complex development transpires among the eosin anion and the protonated analyte via electrostatic and hydrophobic interactions (Fig. 4.14).

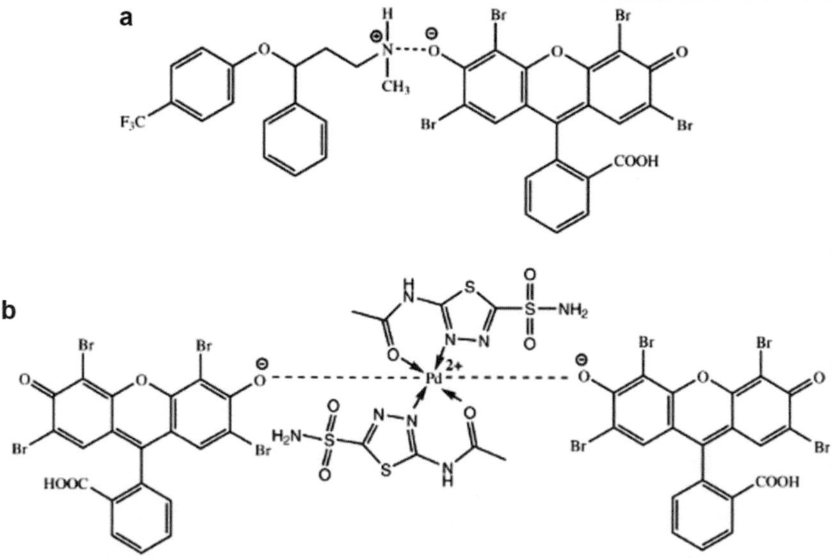

Fig. 4.14 Chemical structures of (A) the binary complex of eosin Y and fluoxetine, and (B) the ternary combination of eosin, palladium, and acetazolamide [68]

4.2.6 Indigo Carmine

Indigo carmine ($C_{16}H_8N_2Na_2O_8S_2$, molecular weight ~ 466.36) is a dark blue powder with chemical structure that has two fundamental groups: a chromophore and a $NaSO_3$ group. $NaSO_3$ imparts solubility in water to the dye, while the chromophore group confers on the substance's distinctive hue (Fig. 4.15). The chromophore group consists of a conjugated system where a C = C bond is substituted by two C = O groups and two NH groups [69–71]. Various derivatives of Indigo carmine are represented in Fig. 4.16.

The food company is a primary domain for the extensive utilization of indigo carmine, recognized as a food ingredient. These substances are extensively utilized in industrialized nations to enhance the flavor, color, fragrance, and texture, as well as to prolong the shelf life of food and drinks. Artificial dyes are favored in the food industry due to several benefits, including protection when exposed to light, oxygen,

Fig. 4.15 Chemical structure of indigo carmine. Reproduced with permission from [72]

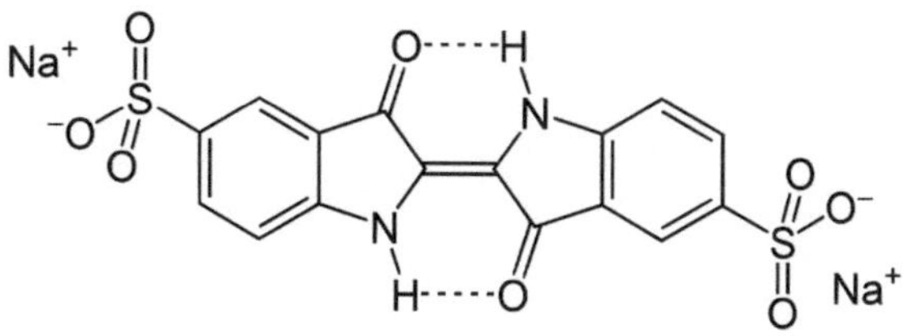

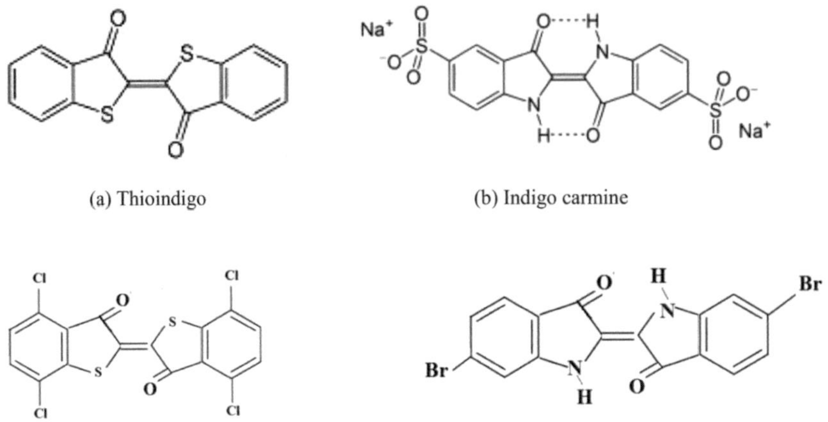

(a) Thioindigo (b) Indigo carmine

(c) 4,4‘,7,7‘-Tetrachlorothioindigo (d) Tyian purple (6,6’-dibromoindigo)

Fig. 4.16 Various derivatives of indigo carmine. Reproduced with permission from [76]

or varying pH levels, uniform coloration of food, excellent solubility, minimal danger of contamination from microbes, and low manufacturing expenses. Prior to the endorsement of a dye in the food sector, extensive toxicity assessments are mandated, including evaluations of both acute and long-term effects, carcinogenicity, persistence, reproductive damage, and bioaccumulation, alongside the determination of the LD50 [73–75]. Indigo carmine was found in textile items from the middle of the nineteenth century. It was primarily utilized for coloring silk, wool, and leather items, but also for coloring polyester and denim fibers. By the conclusion of the nineteenth century, the advancement of chemical synthesis processes had made indigo one of the most significant and expensive substances in the chemical sector [30]. Synthetic indigo in the twentieth century remains a major product of the textile sector, nearly entirely substituting plant-derived indigo. Various derivatives of indigo carmine are shown in Fig. 4.16.

References

1. Nambela, L., Haule, L.V., Mgani, Q.: A review on source, chemistry, green synthesis and application of textile colorants. J. Clean. Prod. **246**, 119036 (2020)
2. Rauf, M., Ashraf, S.S.: Radiation induced degradation of dyes—an overview. J. Hazard. Mater. **166**(1), 6–16 (2009)
3. Pearce, C., Lloyd, J., Guthrie, J.: The removal of colour from textile wastewater using whole bacterial cells: a review. Dyes Pigm. **58**(3), 179–196 (2003)
4. Akpan, U.G., Hameed, B.H.: Parameters affecting the photocatalytic degradation of dyes using TiO2-based photocatalysts: a review. J. Hazard. Mater. **170**(2–3), 520–529 (2009)
5. Fernández, C., Larrechi, M.S., Callao, M.P.: An analytical overview of processes for removing organic dyes from wastewater effluents. TrAC Trends Anal. Chem. **29**(10), 1202–1211 (2010)

6. Rehman, A., et al.: The application of cationic-nonionic mixed micellar media for enhanced solubilization of Direct Brown 2 dye. J. Mol. Liq. **301**, 112408 (2020)
7. Mounia, M.C., et al.: Analyse, synthese et etude solvatochromique de composes tinctoriaux
8. Benkhaya, S., M'rabet, S., El Harfi, A.: A review on classifications, recent synthesis and applications of textile dyes. Inorg. Chem. Commun. **115**, 107891 (2020)
9. Salleh, M.A.M., et al.: Cationic and anionic dye adsorption by agricultural solid wastes: a comprehensive review. Desalination **280**(1–3), 1–13 (2011)
10. Bar, N., Chowdhury, P.: A brief review on advances in rhodamine B based chromic materials and their prospects. ACS Appl. Electron. Mater. **4**(8), 3749–3771 (2022)
11. Kim, H.N., et al.: A new trend in rhodamine-based chemosensors: application of spirolactam ring-opening to sensing ions. Chem. Soc. Rev. **37**(8), 1465–1472 (2008)
12. Chen, X., et al.: Fluorescent chemosensors based on spiroring-opening of xanthenes and related derivatives. Chem. Rev. **112**(3), 1910–1956 (2012)
13. Knauer, K.H., Gleiter, R.: Photochromism of rhodamine derivatives. Angew. Chem. Int. Ed. Engl. **16**(2), 113–113 (1977)
14. Willwohl, H., Wolfrum, J., Gleiter, R.: Kinetics and mechanism of the photochromism of N-phenyl-rhodaminelactame. Laser Chem. **10**(2), 63–72 (1989)
15. Wang, X., Guo, L., Feng, L.: A multi stimuli responsive material with rhodamine B and carbazole groups. New J. Chem. **43**(10), 4181–4187 (2019)
16. Sahu, S., et al.: Kendu (Diospyros melanoxylon Roxb) fruit peel activated carbon—an efficient bioadsorbent for methylene blue dye: equilibrium, kinetic, and thermodynamic study. Environ. Sci. Pollut. Res. **27**(18), 22579–22592 (2020)
17. Amode, J.O., et al.: Adsorption of methylene blue from aqueous solution using untreated and treated (Metroxylon spp.) waste adsorbent: equilibrium and kinetics studies. Int. J. Ind. Chem. **7**(3), 333–345 (2016)
18. Kuang, Y., Zhang, X., Zhou, S.: Adsorption of methylene blue in water onto activated carbon by surfactant modification. Water **12**(2), 587 (2020)
19. Kazemi, F., et al.: Photodegradation of methylene blue with a titanium dioxide/polyacrylamide photocatalyst under sunlight. J. Appl. Polym. Sci. **133**(19) (2016)
20. Lu, Q., Zhang, Y., Liu, S.: Graphene quantum dots enhanced photocatalytic activity of zinc porphyrin toward the degradation of methylene blue under visible-light irradiation. J. Mater. Chem. A **3**(16), 8552–8558 (2015)
21. Huang, S., et al.: Monitoring of oxygen using colorimetric indicator based on graphene/TiO2 composite with first-order kinetics of methylene blue for modified atmosphere packaging. Packag. Technol. Sci. **31**(9), 575–584 (2018)
22. Khan, I., et al.: Review on methylene blue: its properties, uses, toxicity and photodegradation. Water **14**(2), 242 (2022)
23. Dao, H.M., et al.: Methylene blue as a far-red light-mediated photocleavable multifunctional ligand. Chem. Commun. **56**(11), 1673–1676 (2020)
24. Parakala, S., Moulik, S., Sridhar, S.: Effective separation of methylene blue dye from aqueous solutions by integration of micellar enhanced ultrafiltration with vacuum membrane distillation. Chem. Eng. J. **375**, 122015 (2019)
25. Balarak, D., et al.: Application of surfactant-modified bentonite for methylene blue adsorption from aqueous solution. Orient. J. Chem. **36**(2), 293 (2020)
26. Pang, Y., et al.: Effect of humic acid on the degradation of methylene blue by peroxymonosulfate. Open Chem. **16**(1), 401–406 (2018)
27. Abdelrahman, E.A., Hegazey, R., El-Azabawy, R.E.: Efficient removal of methylene blue dye from aqueous media using Fe/Si, Cr/Si, Ni/Si, and Zn/Si amorphous novel adsorbents. J. Mater. Res. Technol. **8**(6), 5301–5313 (2019)
28. Ebi, O.E., Taiwo, A.F., Folorunsho, A.T.: Kinetic modelling of the biosorption of methylene blue onto wild melon (Lagenariasphaerica). Am. J. Chem. Eng. **6**(6), 126–134 (2018)
29. Faizal, A.N.M., Putra, N.R., Zaini, M.A.A.: Insight into the adsorptive mechanisms of methyl violet and reactive orange from water—a short review. Part. Sci. Technol. **41**(5), 730–739 (2023)

30. Hoogenboom, L.: Scientific opinion: malachite green in food. EFSA J. **14**(7), e04530 (2016)
31. Gharavi-Nakhjavani, M.S., et al.: Malachite green and leucomalachite green in fish: a global systematic review and meta-analysis. Environ. Sci. Pollut. Res. **30**(17), 48911–48927 (2023)
32. Srivastava, S., Sinha, R., Roy, D.: Toxicological effects of malachite green. Aquat. Toxicol. **66**(3), 319–329 (2004)
33. Sudova, E., et al.: Negative effects of malachite green and possibilities of its replacement in the treatment of fish eggs and fish: a review. Vet. Med. **52**(12), 527 (2007)
34. Gopinathan, R., Kanhere, J., Banerjee, J.: Effect of malachite green toxicity on non target soil organisms. Chemosphere **120**, 637–644 (2015)
35. Stammati, A., et al.: Effects of malachite green (MG) and its major metabolite, leucomalachite green (LMG), in two human cell lines. Toxicol. In Vitro **19**(7), 853–858 (2005)
36. Au, W., et al.: Cytogenetic toxicity of gentian violet and crystal violet on mammalian cells in vitro. Mutat. Res./Genet. Toxicol. **58**(2–3), 269–276 (1978)
37. Azmi, W., Sani, R.K., Banerjee, U.C.: Biodegradation of triphenylmethane dyes. Enzyme Microb. Technol. **22**(3), 185–191 (1998)
38. Mani, S., Bharagava, R.N.: Exposure to crystal violet, its toxic, genotoxic and carcinogenic effects on environment and its degradation and detoxification for environmental safety. Rev. Environ. Contam. Toxicol. **237**, 71–104 (2016)
39. Ahmad, R.: Studies on adsorption of crystal violet dye from aqueous solution onto coniferous pinus bark powder (CPBP). J. Hazard. Mater. **171**(1–3), 767–773 (2009)
40. Mittal, A., et al.: Adsorption of hazardous dye crystal violet from wastewater by waste materials. J. Colloid Interface Sci. **343**(2), 463–473 (2010)
41. Lillie, R.D., Conn, H.J.: Biological stains: a handbook on the nature and uses of the dyes employed in the biological laboratory. Williams & Wilkins (1969)
42. Luna, L.G.: Manual of histologic staining methods of the Armed Forces Institute of Pathology. In: Manual of histologic staining methods of the Armed Forces Institute of Pathology, pp. xii, 258-xii, 258 (1968)
43. Carson, F.: Histotechnology: A Self-Instructional Text, 2nd edn. ACSP Press, Chicago (1997)
44. Crookham, J.N., Dapson, R.W.: Hazardous chemicals in the histopathology laboratory: regulations risks handling & disposal. Anatech (1991)
45. Churchman, J.W.: The reverse selective bacteriostatic action of acid fuchsin. J. Exp. Med. **37**(1), 1 (1923)
46. Gupta, V., et al.: Adsorption of basic fuchsin using waste materials—bottom ash and deoiled soya—as adsorbents. J. Colloid Interface Sci. **319**(1), 30–39 (2008)
47. Nandi, B.K., Goswami, A., Purkait, M.K.: Adsorption characteristics of brilliant green dye on kaolin. J. Hazard. Mater. **161**(1), 387–395 (2009)
48. Mane, V.S., Mall, I.D., Srivastava, V.C.: Kinetic and equilibrium isotherm studies for the adsorptive removal of Brilliant Green dye from aqueous solution by rice husk ash. J. Environ. Manage. **84**(4), 390–400 (2007)
49. Mane, V.S., Mall, I.D., Srivastava, V.C.: Use of bagasse fly ash as an adsorbent for the removal of brilliant green dye from aqueous solution. Dyes Pigm. **73**(3), 269–278 (2007)
50. Youssef, N.A., et al.: Degradation of methyl orange using Fenton catalytic reaction. Egypt. J. Pet. **25**(3), 317–321 (2016)
51. Qu, W., et al.: Characterization of amino-crosslinked hypromellose and its adsorption characteristics for methyl orange from water. J. Mater. Sci. **55**(17), 7268–7282 (2020)
52. Ali, I., et al.: High-speed and high-capacity removal of methyl orange and malachite green in water using newly developed mesoporous carbon: kinetic and isotherm studies. ACS Omega **4**(21), 19293–19306 (2019)
53. Aljuaid, A., et al.: G-C3N4 based photocatalyst for the efficient photodegradation of toxic methyl orange dye: recent modifications and future perspectives. Molecules **28**(7), 3199 (2023)
54. Kahlert, H., Meyer, G., Albrecht, A.: Colour maps of acid–base titrations with colour indicators: how to choose the appropriate indicator and how to estimate the systematic titration errors. ChemTexts **2**(2), 7 (2016)

55. Yates, E., Yates, A.: Johann Peter Griess FRS (1829–88): Victorian brewer and synthetic dye chemist. Not. Rec. R. Soc. J. Hist. Sci. **70**(1), 65–81 (2016)
56. Siddiqui, S.I., et al.: Investigation of Congo red toxicity towards different living organisms: a review. Processes **11**(3), 807 (2023)
57. Siddiqui, S., et al.: Investigation of Congo red toxicity towards different living organisms: a review. Processes **11**(807), 2023 (2023)
58. Zubair, M., et al.: Enhanced removal of Eriochrome Black T from water using biochar/layered double hydroxide/chitosan hybrid composite: performance evaluation and optimization using BBD-RSM approach. Environ. Res. **209**, 112861 (2022)
59. Alp, H., et al.: Biosorptive removal of Eriochrome Black-T using Agaricus campestris: parameters optimization with response surface methodology. Desalin. Water Treat. **175**, 244–254 (2020)
60. Altaher, H., Khalil, T.E., Abubeah, R.: The effect of dye chemical structure on adsorption on activated carbon: a comparative study. Color. Technol. **130**(3), 205–214 (2014)
61. Moeinpour, F., Alimoradi, A., Kazemi, M.: Efficient removal of Eriochrome black-T from aqueous solution using NiFe2O4 magnetic nanoparticles. J. Environ. Health Sci. Eng. **12**(1), 112 (2014)
62. Abbas, R.F., et al.: Removal of Eriochrome Black T dye by using Al2O3 nanoparticles: central composite design, isotherm and error analysis. Iran. J. Sci. Technol. Trans. A Sci. **44**(4), 993–1000 (2020)
63. Gupta, S.V., Ahmaruzzaman, M.: CeO2/Fe3O4/g-C3N4 nanohybrid for adsorptive removal of Rose Bengal from aqueous stream. Int. J. Environ. Anal. Chem. **104**(7), 1517–1536 (2024)
64. Rauf, M., et al.: Solvatochromic behavior on the absorption and fluorescence spectra of Rose Bengal dye in various solvents. Spectrochim. Acta Part A Mol. Biomol. Spectrosc. **72**(1), 133–137 (2009)
65. Vinuth, M., et al.: Rapid removal of hazardous Rose Bengal dye using Fe (III)–Montmorillonite as an effective adsorbent in aqueous solution. J. Environ. Anal. Toxicol. **6**(2), 1000355 (2016)
66. Poulios, I., et al.: Photooxidation of eosin Y in the presence of semiconducting oxides. Appl. Catal. B **41**(4), 345–355 (2003)
67. Hari, D.P., König, B.: Synthetic applications of eosin Y in photoredox catalysis. Chem. Commun. **50**(51), 6688–6699 (2014)
68. Derayea, S.M., Nagy, D.M.: Application of a xanthene dye, eosin Y, as spectroscopic probe in chemical and pharmaceutical analysis; a review. Rev. Anal. Chem. **37**(3), 20170020 (2018)
69. Kim, I.-S., et al.: Indigo carmine for the selective endoscopic intervertebral nuclectomy. J. Korean Med. Sci. **20**(4), 702–703 (2005)
70. Kim, K.-H., et al.: Time-and dose-dependent cytotoxicities of ioxitalamate and indigocarmine in human nucleus pulposus cells. Spine J. **13**(5), 564–571 (2013)
71. EFSA Panel on Food Additives and Nutrient Sources added to Food (ANS): Scientific opinion on the re-evaluation of Indigo Carmine (E 132) as a food additive. EFSA J. **12**(7), 3768 (2014)
72. Ristea, M.-E., Zarnescu, O.: Indigo carmine: between necessity and concern. J. Xenobiotics **13**(3), 509–528 (2023)
73. Okafor, S.N., et al.: Assessment of the health implications of synthetic and natural food colourants–a critical review. Pharm. Biosci. J. 01–11 (2016)
74. Martynov, V., Brygadyrenko, V.: The influence of the synthetic food colourings tartrazine, allura red and indigo carmine on the body weight of Tenebrio molitor (Coleoptera, Tenebrionidae) larvae. Regul. Mech. Biosyst. **9**(4), 479–484 (2018)
75. Amchova, P., Kotolova, H., Ruda-Kucerova, J.: Health safety issues of synthetic food colorants. Regul. Toxicol. Pharmacol. **73**(3), 914–922 (2015)
76. Ikram, M., Bari, M.A., Haider, J.: Photocatalytic Dye Degradation using Green Polymer-Based Nanostructures: Principles and Applications in Wastewater Treatment. IOP Publishing (2023)

Chapter 5
Techniques for Wastewater Treatment

5.1 Conventional Methods

Numerous causes contribute to water pollution, including homes, industry, mining, and infiltration; however, one of the most significant is the extensive utilization of water by industrial activities [1, 2]. Four types of water are often identified: (1) Rainfall (runoff from impermeable surfaces), (2) household waste, (3) Irrigation water, and (4) wastewater from factories. The last category can be separated into cooling fluid, washing effluent (of varied composition), and production or process water (biodegradable and/or possibly hazardous). Generally, processing waters presents the most significant challenges. Wastewater markedly differs from potable water sources (often rivers, lakes, or reservoirs) in a crucial aspect: contaminant levels in most of the water for consumption sources are far lower than those in effluents originating from industrial activity. Nonetheless, their harmful effects are dependent upon composition, which is influenced by their industrial source. The challenges faced with sewage treatment are inherently complicated, as the effluent comprises diverse contaminants contingent upon its source. Various types of effluents need treatment, each possessing distinct properties that demand particular treatment techniques [3, 4]. A variety of strategies for treating wastewater have been documented in the literature. They generally encompass biological, chemical, and physical procedures that are deemed sufficiently effective for water purification in various manners. Their selection is conditional upon several parameters, including dye content, sewage composition, process cost, and the presence of other contaminants in effluent [3].

M. Ikram et al., *Ferrite Nanomaterials for Wastewater Treatment*, Engineering Materials, https://doi.org/10.1007/978-3-032-09141-3_5

5.2 Physical Methods

The physical method eliminates particles from wastewater using filtration or by extracting solids via gravitational settling or air flotation. Particles trapped in air ascend to the surface and can be extracted. They continue to be the foundation of the majority of methodological flow systems for wastewater treatment [5]. The utilized physical processes are as follows.

5.2.1 Nanofiltration

The concept of nanofiltration (NF) was initially employed in the 1980s to denote membranes exhibiting properties intermediate between ultrafiltration (UF) and the process of reverse osmosis (RO), as shown in Fig. 5.1. It has exhibited significant reliability in several wastewater treatment procedures. It is an effective color removal process that reduces wastewater volume, recycles important components, and recovers them. It operates based on the principles of solution dispersion and pore dimension flow. In NF, the membrane pore dimensions are around 1 nm, resulting in the rejection of even small solutes; yet the surface electrostatic characteristics facilitate the passage of monovalent ions while primarily retaining polyvalent ions. These characteristics render NF membranes highly effective for the fractionation and specific elimination of solutes from intricate processing streams. It is a revolutionary technology for the purification of surface water contaminated by inorganic and organic pollutants [6–8]. The removal of contaminants using NF membranes relies on variations in charge effect and particle dimensions, influenced by electrical and steric forces stemming from Donnan exclusion. A substance with tiny molecules is distinguished by charge effect mechanisms and diffusion, in contrast with the higher molecular weight constituents. Ingredient separation based on the molecular weight using a sieve method.

1. Wetted surface—Water establishes hydrogen bonds with the membrane for transportation
2. Capillary rejection occurs on a microporous and heterogeneous membrane, where the repellent effect arises from the differing electrostatic characteristics of the membrane and the solution.
3. In a solution diffusion process, the membrane is non-porous and homogeneous; both the solute and solvent absorb inside the functional membrane layer, and the movement of the fluid occurs by diffusion across this layer.
4. Charged capillary—rejection based on the pores within the electrical double layer. Owing to streaming possibility, counter ions are repelled while like-charged ions are attracted via the membrane.
5. Finely porous—A membrane consists of solid material with holes. The distinction between pore and bulk fluid governs transit [5].

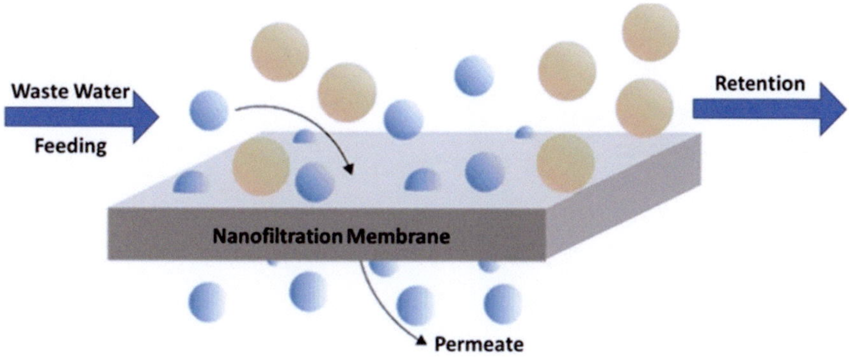

Fig. 5.1 Schematic diagram of NF membrane for water purification. Reproduced with permission from [6]

The characteristics of NF membranes are intermediate between porous ultrafiltration membranes, which reject based on electrical charge effects and dimension exclusion, and non-porous RO membranes, which reject through a solution-diffusion process. It necessitates lower energy consumption and operational expenses compared to RO. NF, in contrast to RO, has superior permeability for water and operates within a small pressure range of 7–30 bar [5].

5.2.2 *Adsorption*

Adsorption is often regarded as an economic and efficient technique for water purification. It is a method of mass transfer wherein solute or biodegradable molecules are conveyed from the liquid phase to the outside of a solid phase (Fig. 5.2). Adsorbed molecules are attached to the solid surface by physicochemical interactions [7–11]. The adsorption removal rate can reach as high as 99.9%. It is regarded as an established method for eliminating colors from polluted water owing to its ease and affordability relative to alternative techniques. The technique of adsorbate movement typically involves three successive steps: (1) the adsorbate movement towards the adsorbent's boundary layer, (2) diffusion (intraparticle) occurs into the pores, and (3) the adsorption as well as desorption of the solute. An adsorbent should ideally possess sufficient binding sites to facilitate effective adsorption of dyes and other contaminants [12]. The predominant traditional adsorbents utilized for the removal of contaminants and dyes include bio-adsorbents, activated carbon (AC), alumina, and titania [13, 14]. AC is regarded as the most effective adsorbent for eliminating various contaminants from polluted water. It is a non-toxic, cost-effective substance characterized by great efficiency due to its porous structure and extensive surface area [10, 15]. Silica (SiO_2) serves as an effective adsorbent substance owing to its

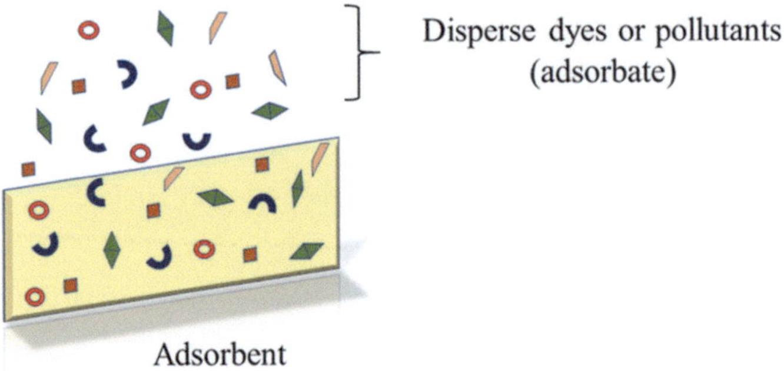

Fig. 5.2 Mechanism of adsorption to remove pollutants. Reproduced with permission from [20]

extensive surface area, consistent pore size, and the possibility of catalytic applications. Mesoporous SiO_2 is regarded as an additive owing to its extensive surface area and favorable permeability for metal oxides. Nano SiO_2 has been efficiently utilized in water purification due to its efficacy, which depends on -OH groups in various contexts [16–18]. Nano Al_2O_3 is an exceptional material for many uses in polluted treatment. However, Al_2O_3 has a restriction in its adsorption capability [19]. Consequently, its adjustments remain in the developmental phase to address the deficit. NMs have garnered significant attention recently as adsorbents in polluted water remediation owing to their extensive specific surface area, reduced flocculent development, and abundant available active sites for pollutant adsorption. Furthermore, these adsorbents are recyclable and reusable, rendering them exceptionally appealing and economical [20].

5.2.3 *Coagulation/Flocculation*

Coagulation is a process wherein insoluble or suspended particulates are aggregated into larger clusters to produce coagulants. Coagulants, like Al, $FeCl_2$, and $FeSO_4$, are often utilized in conventional water purification methods to facilitate the elimination of particulate pollutants by neutralizing particle charges. Figure 5.3 illustrates the application of coagulation-flocculation in the treatment of potable water and the remediation of industrial discharges [21]. It is also efficient for metal ions at quantities below 100 ppm or over 1000 ppm. This technique efficiently eliminates heavy metals when the pH is between 11.0 and 11.5, analogous to precipitation. This procedure cannot completely eliminate pollutants in industrial contaminated, and sludge is generated due to the coagulants employed [22]. Consequently, coagulation must be incorporated with additional contaminated water treatment methods, such as precipitation or chemical reduction. Bojic et al. employed an Al-based composite to examine a combined technique, namely a spontaneous reduction-coagulation procedure for effective removal of metals from effluents. The flow speed, pH, and quantity of metal

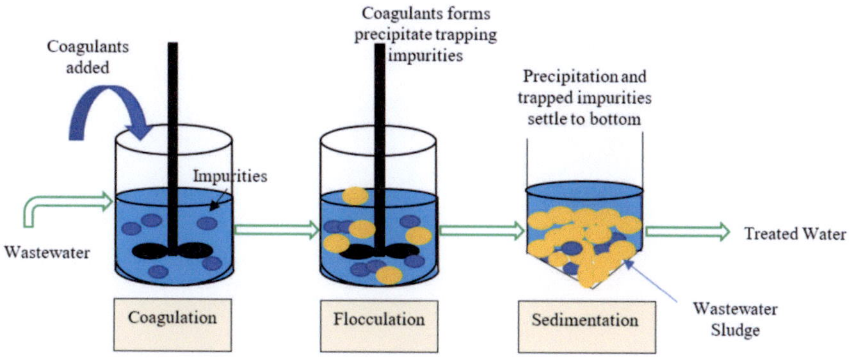

Fig. 5.3 Schematic representation of coagulation/flocculation for water purification. Reproduced with permission from [25]

ions are the factors that are examined to get enhanced elimination capability [23]. Tao et al. investigated the impact of coagulants, such as Ag NPs, on heavy metal content and their ability to decrease pollutants in contaminated water. Coagulation method offers the benefits of rapid settlement of suspended substances and enhanced dehydration capabilities [24]. Conversely, the drawbacks encompass sludge production and elevated operational expenses attributable to chemical necessities.

5.3 Chemical Methods

A variety of chemical techniques have been documented for multiple catalytic purposes. The advanced oxidation process (AOPs) is regarded as a crucial method for water purification. AOPs is an abbreviation denoting all methodologies employed in contaminated water treatment that share similar concepts in the generation of oxidizing species, especially $^{\bullet}OH$. Figure 5.4 illustrates advanced oxidation methods for textile water purification (Fig. 5.4) [26–28].

5.3.1 Advanced Oxidation Process

AOPs were first proposed for the filtration of potable water in the 1980s [30, 31]. AOPs are defined as oxidation processes that generate a substantial quantity of $OH^{\bullet}$, which can effectively cleanse water. The concept of AOP has since been broadened to encompass oxidative processes involving $SO_4^{\bullet-}$. Unlike conventional oxidants such as Cl_2 and O_3, which function as both decontaminants and disinfectants, AOPs are predominantly employed to eradicate organic or inorganic pollutants in wastewater. The eradication of microorganisms and pathogenic indicators by AOP has been

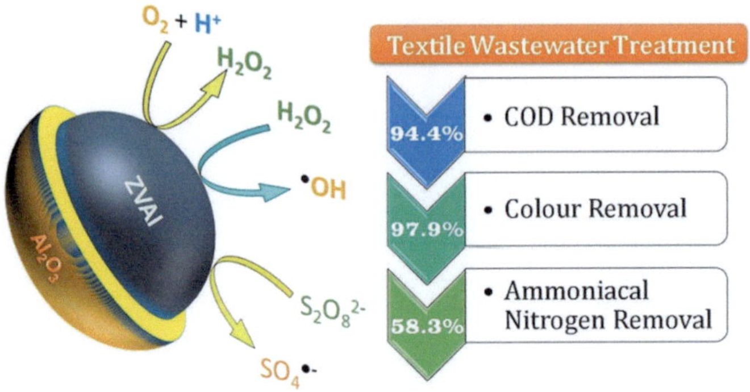

Fig. 5.4 AOPs for water purification. Reproduced with permission from [29]

studied [32, 33], however it is infrequently employed for disinfection because of the brief half-life of the radicals (measured in microseconds). As a result, the required confinement durations for disinfection are unfeasible due to the exceedingly small amount of radicals [34]. In polluted treatment, AOPs employ highly reactive radicals that function as powerful oxidizing agents. It is expected that they will efficiently decompose pollutants in the contaminated and transform them into less hazardous chemicals. AOPs represent a potential and comprehensive solution for water purification [35].

5.3.1.1 UV-Based AOPs

$OH^{\bullet}$ can be generated by photons in the existence of nano-catalysts. TiO_2, Al_2O_3, SnO_2 and ZnO are used as catalysts to generate hole in VB and e^- in CB. The hole and e^- can produce $OH^{\bullet}$· Under UV irradiation, an abundance of oxidants like H_2O_2 or O_3 may generate extra $OH^{\bullet}$ radicals. In addition to that, but at a wavelength is lower than 242 nm, it is also possible for $OH^{\bullet}$ to be formed by the process of photolysis of H_2O (Eqs. 5.1–5.6) [36, 37].

$$TiO_2 + hv \rightarrow e^-_{cb} + hv^+_{vb} \tag{5.1}$$

$$hv^+_{vb} + OH^-_{(surface)} \rightarrow OH^{\bullet} \tag{5.2}$$

$$hv^+_{vb} + H_2O \rightarrow OH^{\bullet} + H^+ \tag{5.3}$$

$$e^-_{cb} + O_{2(absorbed)} \rightarrow O_2^{\bullet -} \tag{5.4}$$

$$H_2O_2 + hv \rightarrow 2OH^{\bullet} \tag{5.5}$$

$$H_2O_2 + hv \rightarrow OH^{\bullet} + H^{\bullet} \quad (5.6)$$

5.3.1.2 Ozone-Based AOPs

O_3 is an energetic oxidizing agent with a 2.07 V anodic potential relative to SCE. O_3 oxidation is a preferential process characterized by typical reaction rate constants of 1.0×10^0–10^3 M^{-1} s^{-1}, wherein O_3 selectively interacts with the ionized and dispersed forms of organic molecules, as opposed to their neutral forms. Under specific circumstances, $OH^{\bullet}$ is generated from O_3 to commence uncontrolled oxidation (indirect processes). Various intricate mechanisms have been suggested to elucidate the complicated creation of $OH^{\bullet}$, with the entire response including $OH^{\bullet}$ generation described as follows (5.7).

$$3O_3 + H_2O \rightarrow OH^{\bullet} + 4O_2 \quad (5.7)$$

$OH^{\bullet}$ yield can be markedly enhanced in the presence of additional oxidants or irradiation. In O_3/H_2O_2 system, the breakdown of O_3 and the formation of $OH^{\bullet}$ are augmented by HO_2^- generated from the breakdown of H_2O_2 (5.8)–(5.9).

$$H_2O_2 \rightarrow HO_2^- + H^+ \quad (5.8)$$

$$HO_2^- + O_3 \rightarrow OH^{\bullet} + O_2^- + O_2 \quad (5.9)$$

During O_3/UV irradiation, H_2O_2 is produced as a supplementary oxidant largely by O_3 photolysis (5.10)–(5.11) [37, 38].

$$O_3 + HO_2^- + hv \rightarrow H_2O_2 + O_2 \quad (5.10)$$

$$H_2O_2 + hv \rightarrow 2OH^{\bullet} \quad (5.11)$$

5.3.1.3 Fenton-Based AOPs

Iron (Fe) is an extremely often utilized metal that activates H_2O_2 and produces $2OH^{\bullet}$ in water. H_2O_2 interacts with Fe^{2+} to form highly reactive substances in the Fenton reaction (5.12)–(5.18) [39].

$$Fe^{2+} + H_2O_2 \rightarrow Fe^{3+} + OH^{\bullet} + OH^- \quad (5.12)$$

$$Fe^{3+} + H_2O_2 \rightarrow Fe^{2+} + HO_2^{\bullet} + H^+ \quad (5.13)$$

$$OH^{\bullet} + H_2O_2 \rightarrow HO_2^{\bullet} + H_2O \tag{5.14}$$

$$OH^{\bullet} + Fe^{2+} \rightarrow Fe^{3+} + OH^{-} \tag{5.15}$$

$$Fe^{3+} + HO_2^{\bullet} \rightarrow Fe^{2+} + O_2H^{+} \tag{5.16}$$

$$Fe^{2+} + HO_2^{\bullet} + H^{+} \rightarrow Fe^{3+} + H_2O_2 \tag{5.17}$$

$$2HO_2^{\bullet} \rightarrow H_2O_2 + O_2 \tag{5.18}$$

H. J. Fenton elucidated the oxidative capacity of H_2O_2 for specific organic substances, whereby $OH^{\bullet}$ are generated from H_2O_2 in the presence of Fe(II) as a catalyst. Subsequently, it was shown that this resulted from the formation of $OH^{\bullet}$, contingent upon the reaction (5.19) [40].

$$Fe^{2+} + H_2O_2 \rightarrow Fe^{3+} + OH^{-} + OH^{\bullet} \tag{5.19}$$

The Fenton reaction is frequently employed due to the ease for the removal of resistant chemicals. The primary disadvantage of the Fenton reaction is the generation of Fe sludge waste, leading to the creation of the photo-Fenton method, that utilizes UV irradiation to reduce Fe(III) oxalate to Fe(II) oxalate, significantly decreasing sludge waste [41].

The breakdown rate of organic contaminants is markedly increased when UV–visible light with wavelengths exceeding 300 nm is introduced to the process (5.20)–(5.22).

$$Fe^{2+} + H_2O_2 \rightarrow Fe^{3+} + OH^{-} + OH^{\bullet} \tag{5.20}$$

$$Fe^{3+} + H_2O \rightarrow Fe^{2+} + H^{+} + OH^{\bullet} \tag{5.21}$$

$$H_2O_2 \rightarrow 2\,OH^{\bullet} \tag{5.22}$$

This method, which utilizes UV light, diminishes the generation of sludge waste created in the original Fenton technique. The reactor size can thus be minimized due to the elevated speed of reaction in this instance. It is essential to meticulously monitor the pH of the medium. The optimal pH range for system efficiency is the range of 2.6 and 3. A solar light-driven photo-Fenton method has been established at the Solar Platform in Almería and was used for the pre-treatment of water employed in the cleaning of fertilizer containers. The Solar Platform asserts that this technique is financially viable, and a comparable facility is now being constructed for the treatment of effluent from the pharmaceutical sector [42].

5.4 Biological Methods

The microbes disintegrate or destroy organic dyes in biological therapies through aerobic or anaerobic processes. Microorganisms utilize organic compounds as a source of power through their degradation. Biofilms are utilized for the elimination of harmful substances from wastewater (Fig. 5.5). Numerous colors utilized in the fabric sector are detrimental to aerobic organisms and contribute to sludge elevation and bulking [43]. The aerobic biological method is deemed inadequate for the decomposition of fabric colors, particularly azo dyes. Additionally, it necessitates a larger area and an extended dynamic retention time. Anaerobic bacteria exhibit modest growth rates, necessitating an extended period for colonization. The anaerobic process for fabric colors produces a greater quantity of hazardous aromatic amines following the decomposition of azo dyes by azoreductase. Aromatic amine generated in an anaerobic state mineralizes into an environmentally benign product upon exposure to air; hence, a combined anaerobic and aerobic approach can be employed for successful treatment [44, 45]. Biofilm strategies are essential in the purification of diverse wastewaters across all biological approaches. Biofilms consist of many microbial populations encased in self-produced extracellular polymeric substance (EPS) matrices [46, 47]. The formation of biofilms comprises five critical phases: (1) Initial adhesion of planktonic organisms to a solid substrate or to each other in a water-based medium; (2) Firm attachment resulting from the production of microorganism-mediated EPS such as poly-hydroxyl groups, which, through H-bonding, anchor microbe to the surface [48]; (3) Formation of monolayer small populations on a solid substrate through either connected or stopped growth due to the proliferation of the original colonizers; (4) Biofilm maturation into a three-dimensional structure by incorporating new planktonic microbes and adhering to detritus from the atmosphere around it; and (5) Diversify or dispersal via either active or passive mechanisms, wherein matrix-encased biofilm cells convert into bacteria by means of a cell-to-cell signaling process [49, 50]. The development of biofilm is influenced by vitamins, pH, temperature, surfaces morphology of support medium, speed, EPS synthesis, and divalent cations [51, 52]. The benefits of biofilm-based contaminated water treatment plants include versatility in operation, reduced spatial requirements, decreased hydraulic absorption time, minimal environmental impact, increased biomass dimensions, and enhanced ability to breakdown complicated compounds [53, 54].

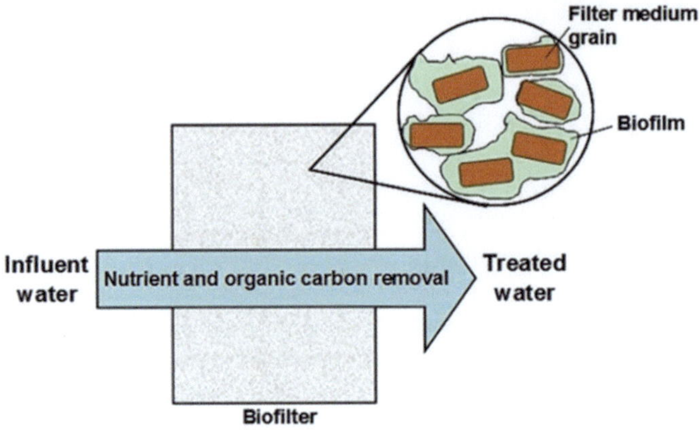

Fig. 5.5 Elimination of contaminants from contaminated using biofilm on filter media. Reproduced with permission from [46]

5.5 Advanced Methods

5.5.1 Catalysis

Catalysis is the process of dye reduction using some reducing agent without the presence of sunlight. NMs including CuO, SnO_2, ZnO, and some others are used as effective catalysts for this process. Sawaira et al. synthesized Mg/chitosan-SnO_2 for effective catalytic MB reduction within very little time. In this process, $NaBH_4$ is employed as reductant and catalyst (Mg/chitosan-SnO_2) helps in the movement of e^- from reductant to MB, resulting in reduction of the dye (Fig. 5.6) [55].

5.6 Photocatalysis

Photocatalytic activity (PCA) is a procedure wherein the rate of chemical transition is accelerated by a material, known as a photocatalyst (PC), under light irradiation without its permanent modification [56]. The striking of light in this procedure develops e^-/h^+ pairs in the CB and VB of the material, which participate in redox reactions to yield the end products [57]. PCA can occur in heterogeneous and homogeneous systems. Homogeneous PCA entails the use of fluid photogenic molecules, such as photoactive dyes, while PCA are expedited by semiconductor PCs in heterogeneous systems. Unlike homogeneous PC, heterogeneous PCs may be easily removed from the reaction media, hence offering a more environmentally friendly approach for commercial use of this system [58, 59]. Heterogeneous PCA possesses significant possibilities for application across various fields, such

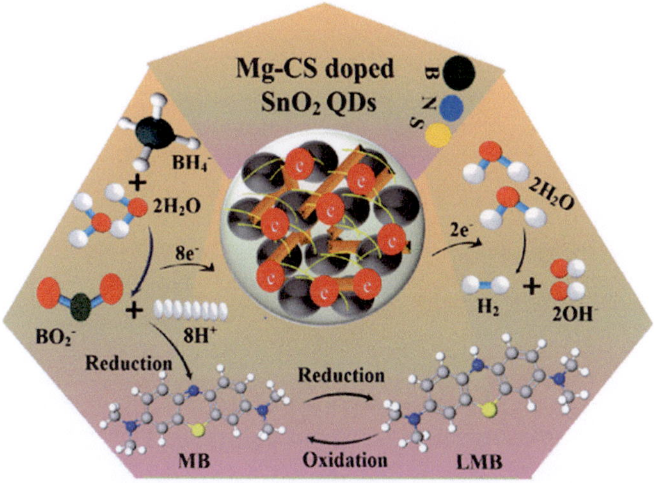

Fig. 5.6 Schematic diagram for catalysis mechanism. Reproduced with permission from [55]

as the elimination of organic contaminants from air and water, decontamination of microorganisms, and H_2 generation through water splitting. PCA is an innovative energy-efficient process that directly transforms light energy into chemical energy, applicable for the breakdown and removal of persistent organic molecules [60, 61]. Notably, PCA offers benefits over traditional procedures owing to its full mineralization of contaminants, ease of use, absence of toxic byproducts, affordability, operation at atmospheric pressure, and reduction of harmful materials (Fig. 5.7).

5.6.1 UV-Light Assisted PCA

UV light occupies the energy spectrum of 10–400 nm. Numerous studies on PCA utilizing a transition metal oxide/UV system have been documented. Dewidar et al. showed that the PCA of ZnO particles in conjunction with UV light was effective in degrading phenol. The decrease of COD for phenol is contingent upon pH, beginning quantity, air surface velocity, and catalyst concentration. The solid–liquid transport of mass was analyzed to comprehend, improve, regulate, and construct the photocatalytic reactor [63]. Saeed et al. synthesized TiO_2/Pd and TiO_2/Pt for usage as PCs in the decomposition investigation of methyl violet dye. It was determined that TiO_2/Pt eliminated 62% of the dye, whereas TiO_2/Pd destroyed around 47% of the dye in an aqueous solution under identical test circumstances (duration: 20 min) [64]. Shao et al. synthesized $ZnFe_2O_4$@ZnO NMs exhibiting enhanced dye reduction by a solvothermal technique. The photocatalytic reduction of MB was achieved under UV irradiation. The core–shell $ZnFe_2O_4$@ZnO has been claimed to have superior reduction activity compared to ZnO [65].

Fig. 5.7 Advantages of PCA for water purification system. Reproduced with permission from [62]

5.6.2 *Visible Light-Assisted PCA*

Typically, just under 5% percent of the total UV light emitted by the sun reaches the Earth's surface. UV energy is inadequate for effective reduction of organic contaminants. Furthermore, UV energy is expensive and poses risks to living things. Visible (VIS) light encompasses a substantial portion of the electromagnetic spectrum. A substance that is active under VIS light is essential for usage and economic viability. Most semiconductors-based PCs are capable of absorbing lower energy VIS light.

Sood et al. produced TiO_2 NMs modified with Bi using the sol–gel technique. The introduction of Bi^{3+} ions into the TiO_2 lattice enhances the reduction potency of Alizarin Red S dye under VIS light. The reduction efficacy was contingent upon many parameters, such as the level of doping of Bi^{3+} ions and the dosage of the synthesized PCs. TiO_2, modified with a 1% of Bi^{3+} ions, exhibited optimal reduction potency for the dye at a dosage of 0.1 g/L. The maximum reduction potency of the dye exceeded 80% within 90 min of response time [66]. Shakira et al. documented the reduction of paracetamol solutions under VIS light irradiation using La-doped ZnO at varied dosages. Observation revealed that 3% La-ZnO NMs demonstrated the best reduction efficacy, achieving 80% decomposition of Paracetamol after 180 min [67]. Patel et al. produced N-TiO_2, V-TiO_2, and V/N-TiO_2 thin films and examined their reduction efficacy on Nitrophenol, MB, and chlorophenol. The V/N-TiO_2 exhibited superior reduction activity relative to the single-doped and pristineTiO_2, attributed to enhanced VIS light absorption and a potential decrease in the recombination of e^--h^+ pairs [68].

5.7 Various Mechanisms of PCA

5.7.1 Conventional Mechanism

The stimulation of e^-s from the VB of NMs to the CB by photons possessing energy exceeding the PCs bandgap energy leads to the creation of h^+ in the VB. The conventional mechanism of PCA depicted as follows (Figs. 5.8, 5.9b, Eqs. 5.23–5.27):

$$\text{Photocatalyst} + \text{hv(UV)} \rightarrow e^-_{cb} + h^+_{vb} \quad (5.23)$$

$$O_2 + e^-_{cb} \rightarrow O_2^- \quad (5.24)$$

$$h^+_{vb} + H_2O \rightarrow {}^\bullet OH \quad (5.25)$$

$$\text{Dye} + h^+_{vb} \rightarrow \text{oxidation product} \quad (5.26)$$

$$\text{Dye} + e^-_{cb} \rightarrow \text{Reduction product} \quad (5.27)$$

5.7.2 Advanced Mechanism

Traditional techniques, including membrane processes and adsorption that primarily regulate water flow while obstructing contaminants, and adsorption, which involves the mechanical adherence of harmful substances to adsorbent surfaces for elimination, are prevalent examples (Fig. 5.9a) [70].

Diverse tactics are utilized to enhance the PCA for potential industrial uses, including doping, the utilization of support, and the development of heterojunctions. The creation of heterojunction PCs enhances the efficacy of each NMs-based PCs by integrating two or more NMs into a single photosystem (PS). The synthesized

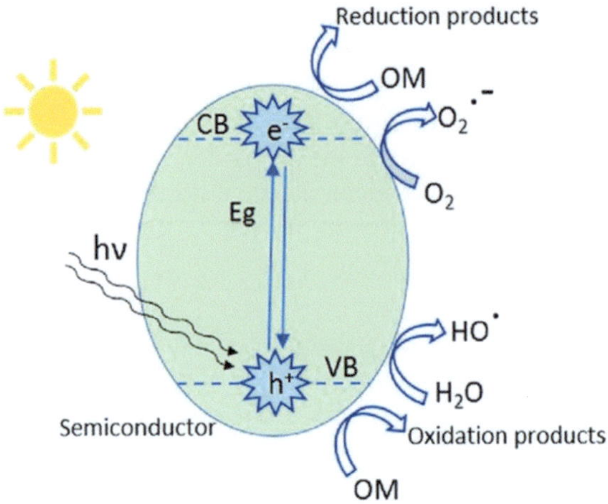

Fig. 5.8 Mechanism of PCA. Reproduced with permission from [69]

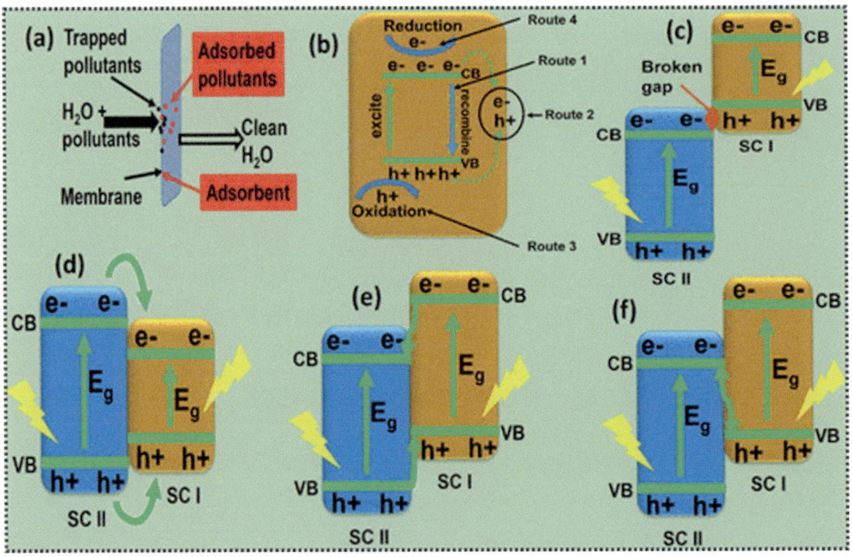

Fig. 5.9 **a** Adsorption process, **b** heterogeneous PCA, and different heterojunctions in PCA: **c** type I, **d** type III, **e** type II, and **f** direct Z-scheme. Reproduced with permission from [76]

heterostructures may be categorized according to the orientation and characteristics of contact and electron transport into type I, type II, type III, and the recently identified Z-scheme heterojunctions [71, 72].

Type I heterojunction, commonly known as the broken gap, features a gap among the VB of PS I and the CB of the PS II, precluding any potential of e^- and h^+

segregation, thereby rendering it inappropriate for PCA (Fig. 5.9c). The type III heterostructure entails charge movement from PS II to PS I, resulting in the accumulation of all e^--h^+ on the VB and CB of PS I, as its edge locations are smaller than those of PS II (Fig. 5.9d). This does not address the swift combination of excited e^- and h^+s, since they accumulate in a single PS with the narrowest Eg, making rapid recombination of e^--h^+ unavoidable. Type II heterojunction is demonstrated to highlight the improved efficiency achieved by integrating two distinct PS (Fig. 5.9e) in contrast to type I and type III systems. This charge movement may be accurately described as the generation of e^--h^+ in PS I and PS II, leading to the movement of e^- from PS I to PS II, while h^+ move in the opposite direction, causing their spatial segregation. In-depth examination of charge movement revealed lowered redox features due to lowering the reaction rates, energy dissipation without direct involvement in PCA, and potential charge repelling between e^- in the CB of PS and h^+ in their corresponding VB during the suggested charge transportation, leading to the emergence of the Z-scheme family. Z-scheme heterojunctions are constructed by combining a pair of PS to deliberately resemble typical photosynthesis, leading to classifications of conventional all-solid-state, and direct Z-scheme according to the utilization of e^- mediators in solution, solid oxidation–reduction mediators, and the absence of mediators, respectively. These heterojunctions offer unique redox sites for redox processes relative to other heterostructures. An examination of the conventional Z-scheme, which utilizes redox PCs alongside fluid redox mediators for interfacial charge movement during recombination, uncovers deficiencies including detrimental redox shuttle side reactions, competition for light absorption between the PCs and redox mediators, and limitations to liquid media applications. The process of the Z-scheme, devoid of redox mediators, was insufficiently elucidated by experimental data upon its initial proposal. Direct Z-scheme PCA employs PCs with appropriate band alignments to facilitate internal electric field-induced recombination of h^+ from the VB of the reductive PS I and e_- from the CB of the oxidative semiconductor, thereby sustaining elevated redox potential and the segregation of e^--h^+ pairs (Fig. 5.9f). The h^+ in the VB of the oxidative PS and the e^- in the CB of the reductive PS are maintained to facilitate redox processes, respectively [73–75].

5.8 Factors Affecting the PCA

5.8.1 *Irradiation Time*

The duration of being exposed to sun light is a key variable influencing the PCA of PCs regarding dye removal [77]. It has been shown that a boost in irradiation duration for the PCA correlates with enhanced photodegradation and removal of color of dye, attributable to the generation of $^{\bullet}OH$ and $^{\bullet}O_2^-$ as irradiation time extends [78]. The deterioration rate first accelerates and thereafter decelerates with prolonged irradiation exposure. Ramli et al. [79] documented fast breakdown of MB dye within the first

30 min, followed by a deceleration in the rate of decomposition thereafter. This is due to the initial production of $^{\bullet}OH$ being more rapid and a greater number of active sites for contaminant adsorption. Following certain reaction duration, the other active sites become challenging to occupy due to the repellent effects among the surface components and the bulk phase, resulting in a tendency for % reduction to stabilize over time [80]. Saeed et al. [81] eliminated MB dye via TiO_2/amidoxime-incorporated polyacrylonitrile, observing that the decomposition rate of the dye escalates with prolonged irradiation duration, ultimately stabilizing after 2 h. The influence of irradiation duration on the breakdown of dye is distinctly illustrated in Fig. 5.10, which depicts image of MO dye before the reaction and following various irradiation times. Figure 5.10 illustrates that an increase in irradiation duration correlates with a rise in the PCA of MO dye, seen by the reduction in color intensity with time.

5.8.2 *Effect of Oxygen Hydroxide*

The existence of H_2O_2 enhances PCA for multiple reasons, such as an elevated quantity of $OH^{\bullet}$, the generation of more oxidative species, a reduction in e^- recombination with h^+, and consequently a boost in the concentration of h^+. The primary factor contributing to the degradation of TiO_2 photocatalysts is the recombination of e^- with positive h^+ (5.28)–(5.30) [83].

$$H_2O_2 + e^- \rightarrow OH^- + OH^{\bullet} \tag{5.28}$$

$$H_2O_2 + O_2^{\bullet} \rightarrow OH^- + OH^{\bullet} + O_2 \tag{5.29}$$

$$H_2O_2 \rightarrow OH^{\bullet} + OH^{\bullet} \tag{5.30}$$

5.8.3 *Effect of Catalyst Dosage*

The quantity of PCs included throughout the PCA is crucial for facilitating dye reduction. The increase in PCs doses lead to an expansion of the overall catalyst surface area. An increased surface area signifies a greater number of active sites present on the PCs surface. In summary, a positive link exists among the dose of the PCs and the deterioration rate. This connection becomes inaccurate when the PCs dose exceed their optimal level. The reaction may be inhibited in the presence of a surplus of PCs. The situation may be elucidated by the obstruction of light absorption into the solution caused by the aggregation of the PCs. As a result, the PCs obstructed and limited the light from illuminating a substantial catalytic region, leading to increased light scattering. The reduction rate decreases when the PCs dose

Fig. 5.10 PCA of MO dye in the existence of **a** Nanoclay **b** CuO/NC **c** CuO [82]

is inadequate. But many PCs may not enhance elimination effectiveness considerably. Elhalil et al. demonstrated that the decomposition efficacy improved with an increase in PCs doses from 0.1 to 0.3 g/L. The maximum caffeine elimination effectiveness was 98.9%. However, a further increase in doses beyond the optimal level will diminish the removal effectiveness because of the screening effect [84–87].

5.8.4 Effect of pH

The pH of the medium for PCA significantly influences the reduction of dyes, as alterations in pH affect PCs surface charge. The PCs acquire a +ve charge with lower pH levels of the medium. Consequently, the anionic dye, owing to its strong Lewis base characteristics, is readily adsorbed onto PCs surface [88]. As the pH of the medium rises, PCs' surfaces typically acquire a -ve charge. The enhanced adsorption of cationic compounds occurs at elevated pH levels. The optimal pH of the solution must be established for various chemical compounds. Bubacz et al. showed that MB reduction, a cationic dye, and phenol escalate with rising pH levels. The optimal results for phenol were obtained at a pH ~ 6.5. The effect of pH on the decolorization of MB dye is ranked as follows: pH 9 > pH 5.8 > pH 3. The maximum photocatalytic degradation of MB was achieved in the basic solution (pH ~ 9), exceeding 96% after 6 h of UV irradiation [89].

Chen et al. 106 documented the influence of pH for crystal violet (CV) reduction utilizing TiO_2 as PCs. PCA was increased as the pH value increased. The cationic CV dye exhibited poor adsorption onto the TiO_2 surface at low pH, thereby resulting in a sluggish PCA of CV [90].

References

1. Anjaneyulu, Y., Sreedhara Chary, N., Samuel Suman Raj, D.: Decolourization of industrial effluents–available methods and emerging technologies–a review. Rev. Environ. Sci. Bio/Technol. **4**(4), 245–273 (2005)
2. Hai, F.I., Yamamoto, K., Fukushi, K.: Hybrid treatment systems for dye wastewater. Crit. Rev. Environ. Sci. Technol. **37**(4), 315–377 (2007)
3. Crini, G., Lichtfouse, E.: Advantages and disadvantages of techniques used for wastewater treatment. Environ. Chem. Lett. **17**(1), 145–155 (2019)
4. Cooney, D.O.: Adsorption Design for Wastewater Treatment. CRC Press (1998)
5. Bahuguna, A., et al.: Physical method of wastewater treatment-a review. J. Res. Environ. Earth Sci **7**, 2348–2532 (2021)
6. Cheryan, M.: Ultrafiltration and Microfiltration Handbook. CRC Press (1998)
7. Anim-Mensah, A.R., Krantz, W.B., Govind, R.: Studies on polymeric nanofiltration-based water softening and the effect of anion properties on the softening process. Eur. Polym. J. **44**(7), 2244–2252 (2008)
8. Tchobanoglous, G.: Fundamentals of biological treatment. In: Wastewater Engineering: Treatment and Reuse, pp. 611–635 (2003)
9. Sharma, A., et al.: Cutting edge technology for wastewater treatment using smart nanomaterials: recent trends and futuristic advancements. Environ. Sci. Pollut. Res. **31**(48), 58263–58293 (2024)
10. Liu, Q., et al.: Novel cyclodextrin-based adsorbents for removing pollutants from wastewater: a critical review. Chemosphere **241**, 125043 (2020)
11. Manchisi, J., et al.: Ironmaking and steelmaking slags as sustainable adsorbents for industrial effluents and wastewater treatment: a critical review of properties, performance, challenges and opportunities. Sustainability **12**(5), 2118 (2020)
12. Labbez, C., et al.: Retention of mineral salts by a polyamide nanofiltration membrane. Sep. Purif. Technol. **30**(1), 47–55 (2003)

13. Wadhawan, S., et al.: Role of nanomaterials as adsorbents in heavy metal ion removal from waste water: a review. J. Water Process Eng. **33**, 101038 (2020)
14. Prajapati, A.K., Das, S., Mondal, M.K.: Exhaustive studies on toxic Cr (VI) removal mechanism from aqueous solution using activated carbon of Aloe vera waste leaves. J. Mol. Liq. **307**, 112956 (2020)
15. Azam, K., et al.: Development of recoverable magnetic mesoporous carbon adsorbent for removal of methyl blue and methyl orange from wastewater. J. Environ. Chem. Eng. **8**(5), 104220 (2020)
16. Slatni, I., et al.: Mesoporous silica synthesized from natural local kaolin as an effective adsorbent for removing of Acid Red 337 and its application in the treatment of real industrial textile effluent. Environ. Sci. Pollut. Res. **27**(31), 38422–38433 (2020)
17. Hussain, M., et al.: Novel mesoporous silica supported ZnO adsorbents for the desulphurization of biogas at low temperatures. Chem. Eng. J. **188**, 222–232 (2012)
18. Ahmed, I.A., et al.: Synthesis and characterization of silica-coated oxyhydroxide aluminum/doped polymer nanocomposites: a comparative study and its application as a sorbent. Molecules **25**(7), 1520 (2020)
19. Kumari, U., et al.: Calcium and zirconium modified acid activated alumina for adsorptive removal of fluoride: performance evaluation, kinetics, isotherm, characterization and industrial wastewater treatment. Adv. Powder Technol. **31**(5), 2045–2060 (2020)
20. Rashid, R., et al.: A state-of-the-art review on wastewater treatment techniques: the effectiveness of adsorption method. Environ. Sci. Pollut. Res. **28**(8), 9050–9066 (2021)
21. Yadav, M., Singh, G., Jadeja, R.: Physical and chemical methods for heavy metal removal. In: Pollutants and Water Management: Resources, Strategies and Scarcity, pp. 377–397 (2021)
22. Chang, Q., Wang, G.: Study on the macromolecular coagulant PEX which traps heavy metals. Chem. Eng. Sci. **62**(17), 4636–4643 (2007)
23. Bojic, A.L., Bojic, D., Andjelkovic, T.: Removal of Cu2+ and Zn2+ from model wastewaters by spontaneous reduction–coagulation process in flow conditions. J. Hazard. Mater. **168**(2–3), 813–819 (2009)
24. Tao, W., et al.: Influence of silver nanoparticles on heavy metals of pore water in contaminated river sediments. Chemosphere **162**, 117–124 (2016)
25. Rajoria, S., Vashishtha, M., Sangal, V.K.: Treatment of electroplating industry wastewater: a review on the various techniques. Environ. Sci. Pollut. Res. **29**(48), 72196–72246 (2022)
26. Shafiq, I., et al.: Recent developments in alumina supported hydrodesulfurization catalysts for the production of sulfur-free refinery products: a technical review. Catal. Rev. **64**(1), 1–86 (2022)
27. Shafiq, I., et al.: Efficient catalyst development for deep aerobic photocatalytic oxidative desulfurization: recent advances, confines, and outlooks. Catal. Rev. **64**(4), 789–834 (2022)
28. Sicardi, S.: Treatment of wastewater from textile dyeing by ozonization (2020)
29. Khatri, J., et al.: Advanced oxidation processes based on zero-valent aluminium for treating textile wastewater. Chem. Eng. J. **348**, 67–73 (2018)
30. Glaze, W.H.: Drinking-water treatment with ozone. Environ. Sci. Technol. **21**(3), 224–230 (1987)
31. Glaze, W.H., Kang, J.-W., Chapin, D.H.: The chemistry of water treatment processes involving ozone, hydrogen peroxide and ultraviolet radiation (1987)
32. Cho, M., et al.: Different inactivation behaviors of MS-2 phage and Escherichia coli in TiO2 photocatalytic disinfection. Appl. Environ. Microbiol. **71**(1), 270–275 (2005)
33. Ikai, H., et al.: Photolysis of hydrogen peroxide, an effective disinfection system via hydroxyl radical formation. Antimicrob. Agents Chemother. **54**(12), 5086–5091 (2010)
34. Yan, S., et al.: Wastewater sludge characteristics. In: Sustainable Sludge Management: Production of Value Added Products, pp. 6–36. American Society of Civil Engineers, USA (2009)
35. Huang, C., Dong, C., Tang, Z.: Advanced chemical oxidation: its present role and potential future in hazardous waste treatment. Waste Manag. **13**(5–7), 361–377 (1993)
36. Tang, W.Z.: Physicochemical Treatment of Hazardous Wastes. CRC Press (2003)

37. Deng, Y., Zhao, R.: Advanced oxidation processes (AOPs) in wastewater treatment. Curr. Pollut. Reports **1**(3), 167–176 (2015)
38. Gottschalk, C., Libra, J.A., Saupe, A.: Ozonation of Water and Waste Water: A Practical Guide to Understanding Ozone and Its Applications. Wiley (2009)
39. Pignatello, J.J., Oliveros, E., MacKay, A.: Advanced oxidation processes for organic contaminant destruction based on the Fenton reaction and related chemistry. Crit. Rev. Environ. Sci. Technol. **36**(1), 1–84 (2006)
40. Walling, C.: Fenton's reagent revisited. Acc. Chem. Res. **8**(4), 125–131 (1975)
41. Vogelpohl, A.: Applications of AOPs in wastewater treatment. Water Sci. Technol. **55**(12), 207–211 (2007)
42. Poyatos, J.M., et al.: Advanced oxidation processes for wastewater treatment: state of the art. Water Air Soil Pollut. **205**(1), 187–204 (2010)
43. Saxena, G., Bharagava, R.N.: Organic and inorganic pollutants in industrial wastes: ecotoxicological effects, health hazards, and bioremediation approaches. In: Environmental Pollutants and Their Bioremediation Approaches, pp. 23–56. CRC Press, Boca Raton (2017)
44. Ibrahim, Z., et al.: Characterisation of microbial flocs formed from raw textile wastewater in aerobic biofilm reactor (ABR). Water Sci. Technol. **60**(3), 683–688 (2009)
45. Yogalakshmi, K.N., et al.: Nano-bioremediation: a new age technology for the treatment of dyes in textile effluents. In: Bioremediation of Industrial Waste for Environmental Safety, vol. I. Industrial Waste and Its Management, pp. 313–347. Springer, Berlin (2019)
46. Sehar, S., Naz, I.: Role of the biofilms in wastewater treatment. Microb. Biofilms-Importance Appl. **10**, 63499 (2016)
47. Naresh Yadav, D., et al.: Evaluation of tire derived rubber (TDR) fixed biofilm reactor (FBR) for remediation of Methylene blue dye from wastewater. Environ. Technol. **42**(23), 3627–3640 (2021)
48. van Belkum, A.: The biofilm mode of life: mechanisms and adaptations. In: Kjelleberg, S., Givskov, M. (eds.) Horizon Bioscience. Elsevier, United Kingdom (2007). ISBN: 978-1-904933-33-5
49. Webb, J.S.: Differentiation and dispersal in biofilms (2009). http://citeseerx.ist.psu.edu/viewdoc/download
50. Paridah, M.T., et al.: We are IntechOpen, the world's leading publisher of Open Access books Built by scientists, for scientists TOP 1%. Intech I **13** (2016)
51. Ansari, F.A., et al.: Factors affecting biofilm formation in in vitro and in the rhizosphere. Biofilms Plant Soil Health 275–290 (2017)
52. Krivorot, M., et al.: Factors affecting biofilm formation and biofouling in membrane distillation of seawater. J. Membr. Sci. **376**(1–2), 15–24 (2011)
53. Martin, K.J., Nerenberg, R.: The membrane biofilm reactor (MBfR) for water and wastewater treatment: principles, applications, and recent developments. Bioresour. Technol. **122**, 83–94 (2012)
54. Lewandowski, Z., Boltz, J.: Biofilms in water and wastewater treatment. In: Water-Quality Engineering, pp. 529–570. Elsevier (2011)
55. Moeen, S., et al.: Comparative study of sonophotocatalytic, photocatalytic, and catalytic activities of magnesium and chitosan-doped tin oxide quantum dots. ACS Omega **7**(50), 46428–46439 (2022)
56. Nath, R.K., Zain, M., Jamil, M.: An environment-friendly solution for indoor air purification by using renewable photocatalysts in concrete: a review. Renew. Sustain. Energy Rev. **62**, 1184–1194 (2016)
57. Moniz, S.J., et al.: Visible-light driven heterojunction photocatalysts for water splitting–a critical review. Energy Environ. Sci. **8**(3), 731–759 (2015)
58. Friedmann, D., et al.: Heterogeneous photocatalytic organic synthesis: state-of-the-art and future perspectives. Green Chem. **18**(20), 5391–5411 (2016)
59. Saadati, F., Keramati, N., Ghazi, M.M.: Influence of parameters on the photocatalytic degradation of tetracycline in wastewater: a review. Crit. Rev. Environ. Sci. Technol. **46**(8), 757–782 (2016)

60. Chen, J., et al.: The application of heterogeneous visible light photocatalysts in organic synthesis. Catal. Sci. Technol. **6**(2), 349–362 (2016)
61. Keller, K.M., et al.: 20th century changes in carbon isotopes and water-use efficiency: tree-ring-based evaluation of the CLM4. 5 and LPX-Bern models. Biogeosciences **14**(10), 2641–2673 (2017)
62. Pirhashemi, M., Habibi-Yangjeh, A., Pouran, S.R.: Review on the criteria anticipated for the fabrication of highly efficient ZnO-based visible-light-driven photocatalysts. J. Ind. Eng. Chem. **62**, 1–25 (2018)
63. Dewidar, H., Nosier, S., El-Shazly, A.: Photocatalytic degradation of phenol solution using Zinc Oxide/UV. J. Chem. Health Saf. **25**(1), 2–11 (2018)
64. Saeed, K., et al.: Efficient photodegradation of methyl violet dye using TiO2/Pt and TiO2/Pd photocatalysts. Appl Water Sci **7**(7), 3841–3848 (2017)
65. Shao, R., et al.: Preparation and characterization of magnetic core–shell ZnFe2O4@ ZnO nanoparticles and their application for the photodegradation of methylene blue. Chem. Eng. J. **217**, 185–191 (2013)
66. Sood, S., et al.: The visible light-driven photocatalytic degradation of Alizarin red S using Bi-doped TiO 2 nanoparticles. New J. Chem. **38**(7), 3127–3136 (2014)
67. Shakir, M., et al.: Photocatalytic degradation of the Paracetamol drug using Lanthanum doped ZnO nanoparticles and their in-vitro cytotoxicity assay. J. Lumin. **176**, 159–167 (2016)
68. Patel, N., et al.: Efficient photocatalytic degradation of organic water pollutants using V-N-codoped TiO2 thin films. Appl. Catal. B **150**, 74–81 (2014)
69. Palani, G., et al.: Metal-doped graphitic carbon nitride nanomaterials for photocatalytic environmental applications—a review. Nanomaterials **12**(10), 1754 (2022)
70. Huang, R., et al.: A study on the adsorption behaviors of three hydrophobic quinolones by ordered mesoporous CMK-3. Chemosphere **294**, 133761 (2022)
71. Malefane, M., Feleni, U., Kuvarega, A.: Cobalt (II/III) oxide and tungsten (VI) oxide pn heterojunction photocatalyst for photodegradation of diclofenac sodium under visible light. J. Environ. Chem. Eng. **8**(2), 103560 (2020)
72. Paul, T., et al.: CsPbBrCl2/g-C3N4 type II heterojunction as efficient visible range photocatalyst. J. Hazard. Mater. **380**, 120855 (2019)
73. Zhang, G., Wang, Z., Wu, J.: Construction of a Z-scheme heterojunction for high-efficiency visible-light-driven photocatalytic CO_2 reduction. Nanoscale **13**(8), 4359–4389 (2021)
74. Schumacher, L., Marschall, R.: Recent advances in semiconductor heterojunctions and Z-schemes for photocatalytic hydrogen generation. Top. Curr. Chem. **380**(6), 53 (2022)
75. Low, J., et al.: A review of direct Z-scheme photocatalysts. Small Methods **1**(5), 1700080 (2017)
76. Malefane, M.E., et al.: Understanding the principles and applications of dual Z-scheme heterojunctions: how far can we go? J. Phys. Chem. Lett. **14**(4), 1029–1045 (2023)
77. Makeswari, M., Saraswathi, P.: Photo catalytic degradation of methylene blue and methyl orange from aqueous solution using solar light onto chitosan bi-metal oxide composite. SN Appl. Sci. **2**(3), 336 (2020)
78. Sakthivel, S., et al.: Solar photocatalytic degradation of azo dye: comparison of photocatalytic efficiency of ZnO and TiO2. Sol. Energy Mater. Sol. Cells **77**(1), 65–82 (2003)
79. Che Ramli, Z.A., et al.: Photocatalytic degradation of methylene blue under UV light irradiation on prepared carbonaceous TiO2. Sci. World J. **2014**(1), 415136 (2014)
80. Aziztyana, A.P., et al.: Optimisation of methyl orange photodegradation using TiO2-zeolite photocatalyst and H2O2 in acid condition. IOP Conf. Ser. Mater. Sci. Eng. IOP Publishing (2019)
81. Saeed, K., Khan, I., Park, S.-Y.: TiO2/amidoxime-modified polyacrylonitrile nanofibers and its application for the photodegradation of methyl blue in aqueous medium. Desalin. Water Treat. **54**(11), 3146–3151 (2015)
82. Khan, I., et al.: Nanoclay-mediated photocatalytic activity enhancement of copper oxide nanoparticles for enhanced methyl orange photodegradation. J. Mater. Sci. Mater. Electron. **31**(11), 8971–8985 (2020)

83. Al-Nuaim, M.A., Alwasiti, A.A., Shnain, Z.Y.: The photocatalytic process in the treatment of polluted water. Chem. Pap. **77**(2), 677–701 (2023)
84. Abdel-Khalek, A.A., Mahmoud, S., Zaki, A.: Visible light assisted photocatalytic degradation of crystal violet, bromophenol blue and eosin Y dyes using AgBr-ZnO nanocomposite. Environ. Nanotechnol. Monit. Manag. **9**, 164–173 (2018)
85. Saravanan, R., Gracia, F., Stephen, A.: Basic principles, mechanism, and challenges of photocatalysis. In: Nanocomposites for Visible Light-Induced Photocatalysis, pp. 19–40. Springer, Berlin (2017)
86. Elhalil, A., et al.: Photocatalytic degradation of caffeine as a model pharmaceutical pollutant on Mg doped ZnO-Al2O3 heterostructure. Environ. Nanotechnol. Monit. Manag. **10**, 63–72 (2018)
87. Koe, W.S., et al.: An overview of photocatalytic degradation: photocatalysts, mechanisms, and development of photocatalytic membrane. Environ. Sci. Pollut. Res. **27**(3), 2522–2565 (2020)
88. Li, F., et al.: Photodegradation of an azo dye using immobilized nanoparticles of TiO2 supported by natural porous mineral. J. Hazard. Mater. **152**(3), 1037–1044 (2008)
89. Bubacz, K., et al.: Methylene blue and phenol photocatalytic degradation on nanoparticles of anatase TiO 2. Pol. J. Environ. Stud. **19**(4) (2010)
90. Chen, C.-C., et al.: Photocatalyzed N-de-methylation and degradation of crystal violet in titania dispersions under UV irradiation. Dyes Pigm. **75**(2), 434–442 (2007)

Chapter 6
Conclusion, Challenges, and Future Perspective

Anticipated population growth and changing consumption patterns are likely to increase global freshwater demands beyond sustainable thresholds in the future, especially in the semi-arid and arid regions of the Middle East and North Africa (MENA), Southeast Asia, and Australia. From a Pan-European viewpoint, southern regions, particularly those in the Mediterranean, are markedly vulnerable to imminent freshwater shortages. The escalating temporal and spatial unpredictability of freshwater supplies due to climate change will intensify this scarcity through seasonal droughts. Water use is anticipated to rise by 50% by 2030. Presently, two-thirds of the world's population endures water scarcity for a minimum of one month each year. By 2025, the situation is expected to deteriorate, perhaps leading to half of the population being underwater stress. Although water covers 71% of the Earth's surface, 97% of this volume consists of salt water, rendering it unsuitable for direct intake or daily use. Various water treatment systems outlined at the end of this book differ from one another regarding their foundational ideas, applicable ranges, processing rates the operational context and associated expenses. Adsorption is regarded as the most adaptable and efficient technique, capable of removing a broad spectrum of contaminants, both inorganic and organic. When all these factors are considered, adsorption becomes the clear frontrunner. Furthermore, it is an expedient technique with little expenditure related to its creation, maintenance, and operation. Despite the considerable expense associated with installation and maintenance, reverse osmosis is a method that has extensive use and popularity owing to the superior quality of water it generates. Other methods, including nanofiltration, biological and ion exchange, are employed for specific purposes; however, their usage in wastewater purification on a larger scale is limited. Traditional treatment methods cannot completely eradicate organic contaminants from wastewater. Conversely, these procedures generate secondary pollutants that can frequently exhibit toxicity levels much above those of primary pollutants. These secondary contaminants need further treatment, hence increasing the total cost of the procedure. Photocatalysis is a suitable approach for wastewater treatment as it degrades toxic substances, organic compounds, pesticides,

M. Ikram et al., *Ferrite Nanomaterials for Wastewater Treatment*, Engineering Materials, https://doi.org/10.1007/978-3-032-09141-3_6

dyes, and bacteria and produces no secondary pollutants. Various nanomaterials, such as TiO_2, ZnO, CuO, SnO_2, and so on are used as photocatalysts for wastewater treatment. In this book, we discuss ferrites for wastewater treatments, as many ferrites has a narrower bandgap energy than TiO_2, so they can absorb maximum visible light. Due to their magnetic properties, they are easily recycled and probability to reduce secondary pollutants. Ferrites are stable at various pH levels and provide a larger surface area for photocatalysis. Doping of ferrite with other metals and polymers enhanced their activity and stability for photocatalysis.

Challenges

Despite their potential, many significant hurdles persist that obstruct the extensive implementation of ferrite nanomaterials:

- Electron–hole pair recombination
- High bandgap energy in some ferrite
- Low surface area in bulk ferrite
- Nanoparticle aggregation
- Scalability and costs

Suggestion

- Doping and composite formation reduce the bandgap energy and recombination rate.
- Modifying the properties of ferrite with 2D materials as graphene and graphitic carbon nitride will enhance their conductivity.
- Analysis on an industrial level to verify stability and recyclability.
- Development of an environment-friendly and cost-effective process for their synthesis.

MIX
Papier aus verantwortungsvollen Quellen
Paper from responsible sources
FSC® C105338

If you have any concerns about our products,
you can contact us on
ProductSafety@springernature.com

In case Publisher is established outside the EU,
the EU authorized representative is:
Springer Nature Customer Service Center GmbH
Europaplatz 3, 69115 Heidelberg, Germany

Printed by Libri Plureos GmbH
in Hamburg, Germany